MW00611542

The Dirty Side of the Garment Industry

Fast Fashion and Its Negative Impact on Environment and Society

The Dirty Side of the Garment Industry

Fast Fashion and Its Negative Impact on Environment and Society

Nikolay Anguelov

CRC Press
Taylor & Francis Group
Boca Raton London New York

CRC Press is an imprint of the
Taylor & Francis Group, an **Informa** business

CRC Press
Taylor & Francis Group
6000 Broken Sound Parkway NW, Suite 300
Boca Raton, FL 33487-2742

© 2016 by Taylor & Francis Group, LLC
CRC Press is an imprint of Taylor & Francis Group, an Informa business

No claim to original U.S. Government works

Version Date: 20150717

International Standard Book Number-13: 978-1-4987-1222-4 (Hardback)

Visit the Taylor & Francis Web site at
http://www.taylorandfrancis.com

and the CRC Press Web site at
http://www.crcpress.com

Contents

Acknowledgments

The writing of this book was a process of uncovering information not easily found in research sources. The inside knowledge of fashion industry professionals and economic development policy analysts directed the quest to identify, synthesize, analyze, and link together the varied academic disciplines that examine the global commerce of apparel.

With respect to specific guidance in fashion-related international trade dynamics, especially helpful was the input from Professor Patrick Yanez and Dr. Kristine Pomeranz of the International Trade and Marketing Department at the Fashion Institute of Technology as well as the legal direction and input of Sophie Miyashiro, US Customs Broker. Direction in legal issue analysis also came from Professor Chad McGuire, LL.M., Chair of the Department of Public Policy at the University of Massachusetts, Dartmouth.

I am especially grateful for the ongoing support and training in international economics I received under the tutelage of two World Bank economists who have devoted their careers to issues of global justice and poverty alleviation in the developing world—Dr. Holley Ulbrich and Dr. William A. Ward. I am thankful to both for their input on neoclassical and neo-institutional economics theory, which helped define the contents of Chapters 6 and 7. I must express a special note of gratitude to Dr. Ward for his ongoing support of the project and continuous input with valuable insights on issues of developing nation industrial upgrading and global governance. In addition, I am thankful for the enthusiastic encouragement and direction from Dr. Aleda Roth to focus on the gaps in the supply chain management literature with respect to the global carbon footprint of international fashion economics.

I am also grateful to Joshua Botvin for his editorial help and assistance with manuscript layout and preparation.

Introduction

The garment industry is one of the largest, most globalized, and most essential industries in the modern world (Jansson and Power, 2010). It is among the most global industries, because most nations produce clothing components not only for domestic consumption but for the entire international textile and apparel market (Gereffi and Frederick, 2010). It is arguably one of the most essential industries because of its legacy in national industrial upgrading. The growth of clothing exports has been among the starter industrial policies for countries climbing the industrialization ladder because of its low fixed costs and emphasis on labor-intensive manufacturing (Akamatsu, 1961, 1962; Gereffi, 1999; Gereffi and Drederick, 2010; Kumagai, 2008).

In recent years, technological advances in production, societal changes in wealth, growth due to globalization, and market changes at the retail level have increased its ecological impact (Birnbaum, 2005, 2008; Kunz and Garner, 2011). Technologically, fiber quality has been improved, bringing about an increasing variety of natural, synthetic, and blended fabrics. Often referred to as poly blends, these fabrics are ubiquitous because of their durability, quality, and versatility. Improvements in technological efficiency in manufacturing have changed the pricing of fabrics in a downward trend, whereby higher quality is available at ever-decreasing prices. This fact has spurred the advent and growth of "fast fashion." Fast fashion erased the separation of garment strata. Historically, industry insiders referred to "fashion" and "garments" separately. Clothes described as "fashion" denoted higher price and target markets, while "garments" denoted lower price and mass markets. The price differential was tied to design creativity, but more importantly, to the garment quality. Higher quality was defined by fabric characteristics such as finery, color, pattern, and type. Expensive clothes were made from expensive fabrics.

Today, thanks to the abovementioned technological advances, with very few exceptions such as silks, wools, and furs, there is no such thing as an expensive fabric. This reality is best illustrated with an example from the popular American TV show *Project Runway*. In the show, competitors in the field of high fashion design create garments on a fabric budget of $150 (or less). The judges laud the contestants when their creations look expensive while staying on or, even better, under budget. In many cases, these creations, made from the same fabrics, are put directly into the retail stratosphere in an effort to promote new designers as well as make fashionable clothes affordable and accessible. Sounding noble and promising, this idea of bringing the creative force of fashion into the forefront of social attention sends a message. The message is that in order to succeed, an industry entity—be it a designer, a retailer, a merchandizer, or, as is often the case, the combined force of all three—must compete on price. The ability to bring ever-increasing quality and ever-decreasing prices is the ticket to fashion success.

On the surface, this message is good. It offers consumers value. It promises an easier way into the industry for budding designers who can make their creations on low budgets. It allows these creations to be noticed fast and get fast input, creating a better learning curve for designers who can keep on trying to make it, even after a badly received collection or two. It offers hope for future artists who want to explore careers in the sector by providing the ability to work with many entrepreneurial up-and-comers. It increases the incentives for entrepreneurial activity across sectors that can support not just clothes production, marketing, and distribution, but also the production of accessories, shoes, cosmetics, home products, and even pharmaceutical goods devoted to a fashionable lifestyle.

As a result of this proliferation, fashion is no longer a status universe reserved for the select few who can afford it. It is no longer created by a select few who have earned the right to dictate it. Fashion is now "fast fashion"—rapidly available to many, created by many, promoted by many, enjoyed by all. All this new fashion reality is based on affordability—an affordability built on cheap fabrics.

This affordability carries an uncomfortable secret. The production of those inexpensive but high-quality fabrics that enable the diffusion of fashion to the masses creates more toxic chemical pollution per item than any other industrial product. According to World Bank estimates, globally 20% of all water pollution is created during the runoff processes of textile dyeing and rinsing of natural (mostly cotton) fabrics. This estimate does not include the other two main processes in natural fiber treatment: mercerizing (dipping the fibers in a hydrogen peroxide bath to make them more pliable) and bleaching, which needs no explanation. Cotton, and other natural fiber yarns such as linen, are bleached to diminish their naturally yellowish tint to neutral in the preparation for the multiple dyeing and fiber-blending processes. This fact means that on top of the dyes that run off into open-water basins during dyeing and rinsing (which, in modern textile treatment methods, occurs several times per batch of textile), tons of industrial-grade peroxide and bleach are also expelled as effluents into open-water basins. Industrial-grade chemicals have higher concentrations than their commercially sold counterparts.

The combined ecological impact of the above-noted processes makes textile production, arguably, the most polluting industrial sector that produces goods for mass consumption. In support of that claim, there is little empirical evidence allowing comparative analysis with other industries. As is, current estimates indicate that textile production is second only to agricultural production in total freshwater pollution. And yet, unlike in agriculture, little is done to change this alarming reality. In agricultural production, examples of methods for environmental stewardship abound in watershed management and in the disciplines that focus on water purification for agrarian use. In textile production, however, pollution is simply noted as a problem, and that is where the discussion ends. In the context of the entire industry, this fact is worrisome, because garment industry analysis seldom includes ecological components. The analysis is focused on the increase of productivity associated with the progress in developing reactive and responsive business models in fashion economics as a result of the proliferation of fast fashion. Supply chain and production management extol fast fashion's merits, from responsiveness, to agility, to profitability. Agroeconomists commend the productivity associated with cotton exports. Trade professionals praise the increasing volumes of international trade in fashion and related industries and the wealth this trade is creating. Consequently, garment production and sales are at all-time highs, spurred by growing demand and consumption. This book questions the ethics of the promotion of this ever-increasing magnitude of consumption.

Author

Nikolay Anguelov is a professor of economic development in the Department of Public Policy at the University of Massachusetts, Dartmouth. He oversees the Public Management concentration in the Master of Public Policy (MPP) program and teaches courses in Statistical and Policy Analysis, Economic Development, and Globalization. Dr. Anguelov is an interdisciplinary scholar whose work focuses on the meeting points between economics, politics, and diplomacy. His research focus is a product of his private sector experience in international trade. As an alumnus of the Fashion Institute of Technology, Dr. Anguelov started his career in the private sector of fashion and home product international commerce, in which he eventually started his own successful business. These experiences define his attention to the role policy plays in creating economic incentives both domestically and internationally. Dr. Anguelov is also the author of *Policy and Political Theory in Trade Practice: Multinational Corporations and Global Governments* (2014) and *Economic Sanctions vs. Soft Power: Lessons from Myanmar, North Korea and the Middle East* (2015), both published by Palgrave Macmillan.

1

From Fashion to Fast Fashion

A NEW WAY TO CREATE TRENDS

Among the many industrial sectors most changed by globalization, the fashion and related industries stand out. Fashion and related industries is a collective classification that refers to the garment, footwear, accessory, cosmetics, and fragrance sectors. Often, for simplicity and inclusivity, this amalgam of sectors is described as the apparel industry. Today, it is the most pervasive and internationalized industry in the world. It encompasses all facets of dressing, from underwear to outerwear, to shoes, bags, hats, belts, gloves, and other accessories, to jewelry and makeup, fragrance, and bath products, to sports gear and adventure wear, to work and military gear. All these facets of dressing change based on trends. This book looks at the setters and promoters of trends and their impact on the general public's spending habits in the quest to be fashionable.

Trends are created. Their creation and destruction are said to be outcomes of cultural change. But culture is slow to change, while trends are not. Culture is remarkably stable at its core because it relies on the stability of social norms (Montzavinos, 2001). Therefore, the overall conclusion that a change in culture causes a change in fashion trends does not account for the speed with which trends change today.

Trends change and have always changed. What is different today is that their modern rate of change is much higher than it was two short decades ago. It is not just higher; it is exponentially higher. Therefore, it is more accurate to say that trends are changed, rather than that trends change. In fashion, that change is not cultural, nor is it unplanned and organic; rather, it is calculated. Trends are introduced for fashion seasons in order to stimulate sales.

In the past, industry platforms in the fashion world had developed around eight traditional "seasons"—Spring, Summer I, Summer II, Fall, Trans-seasonal, Winter I, Winter II, and Holiday (Birnbaum, 2005). Trend modifications were offered in each season, and fashion-forward customers eagerly anticipated the latest trend. Today, 24 different distinct "seasons" exist, including such new classifications as "back to school," "prom," and even "wedding"—a separate late spring "season" for when most young people purchase special outfits for weddings, which tend to be popularly scheduled at that time of the year (Christopher et al., 2004; Gaimster, 2012; Jimenez and Kolsun, 2014; Nenni et al., 2013).

It could be cultural factors that have contributed to the granulation of the modern fashion trend-setting cycle. The question is determining what factors elicited the supposed cultural change that necessitates the addition of so many new trend seasons. It is an important question, because it has changed the consumption and, therefore, production volumes of the industry. Consumers around the world are buying a lot more clothing (Brans and Lea-Greenway, 2006). Few, if any, other consumer good industrial sectors have grown to levels of aggregate consumption comparable to clothes.

Clothing as a whole simultaneously covers the spectrum from a staple product that can be seen as a necessity to a luxury product that can be seen as a stamp of high social standing. Where a particular garment lies in that spectrum is a matter of perception based on trends. A look into the stratification of that spectrum is warranted because the distance, so to speak, between the lowest-end garments (the necessity product) and the highest-priced fashionable garments (the luxury product) has decreased. The decrease is related to the shrinkage of the life span of trends.

Historically, clothes were defined easily into low-end, mass-market, and high-end items separated clearly by price, brand, and quality; today, the difference between a $10 skirt and a $200 skirt can be undetectable. Items that look high-end in terms of design, brand, and quality can be sold for less than the price of a sandwich. As a result, in many places, cheap, readily disposable clothes have displaced hand-me-downs or more durable garments as the mainstay of dressing (Rosenthal, 2007).

This reality of low prices carries its own set of costs. In order to keep prices low for the consumer, clothing manufacturers strive to keep their production costs low. This quest to minimize direct costs has resulted in increases in associated social costs. Social costs are costs borne by society that cannot be internalized by formal markets (Kapp, 1950). Examples are the costs of pollution from industrial production; the costs of social unrest from labor exploitation, if that unrest leads to violence or significant human suffering; the costs of human health damage from side effects of the consumption of unsafe products; and the list today grows longer and longer. With respect to social costs from fashion production, in the recent past, garment conglomerates have had to respond to charges on issues ranging from exploitative labor practices, to perpetuating global poverty by fighting against wage increases in developing nations, to pursuing an unethical marketing strategy of promoting the overconsumption of cheap, readily disposable clothes and therefore contributing to high levels of industrial waste (Diebacker, 2000; Nimon and Beghin, 1999; Rosenthal, 2007).

The production and promotion of such cheap and readily disposable clothes are a recent phenomenon referred to as *fast fashion* (Barn and Lea-Greenway, 2006; Bhardwaj and Fairhurst, 2010; Birnbaum, 2005, 2008; Bruce and Daly, 2004, 2006). The retailers reshaping the industry are European conglomerates such as Zara, H&M, and New Look, which offer inexpensive merchandise that looks expensive (Rosenthal, 2007). Their successful business model has inspired all retailers to explore fast-fashion product lines. Even traditional luxury brands such as Versace now offer fast-fashion options in search of reaching a larger market demographic. These are options that are priced low and therefore can be afforded by customers who were traditionally priced out, so to speak, from owning a luxury brand. It is this low-price quest that fast-fashion retailers successfully deployed, and, as this book outlines, revolutionized the industry.

There exists a complicated decision process of price positioning in which clothing producers engage in order to determine whether consumers will spend $10, $200, or some intermediate amount on a garment. This game, so to speak, was driven by the ability of producers to create and promote trends. The ability to push forward particular trends successfully defined the producer brands, created brand equity and prestige, and enabled charging higher prices. As trends waned, so did the prices. This cycle created different strata of consumers, whereby those most interested in being fashion forward paid the highest price to be the most fashionable. But things have changed.

Technological and industry-specific innovations have led to a new reality, in which consumers find themselves with an increasing number of fashionable choices that do not demand a high price. Today, a new crop of retailers offer fashion-forward choices at price points so low that a new competition has emerged, redefining old market classifications. The line between low end, mass market, and luxury is blurred. That line was previously defined by price. When price is removed from the differentiating equation, retailers readjust their pricing strategies in a downward spiral, always competing to offer the highest-end product at the lowest price point.

When it comes to choosing clothes, consumers have options that allow them to purchase garments and related goods either as daily-life must-haves or as luxury items. Historically, there were two markets to denote those options—a mass market and a luxury market. But today, there are intermediate options. There is the *prestige* market—in which garments, cosmetics, and accessories are priced just below luxury levels. There is also the *mastige* market—closer in price to mass-market items, although slightly higher, but closer in taste and style, and even quality, to luxury items.

The fashion industry was developed around the pathway of elevating garments into staples indicating social status, taste level, general cultural awareness, and personal individuality. This elevation was, and still is, dictated by a change in the latest trends. The change of the longstanding definition of "latest trends" is at the core of the advent of fast fashion. Trends have historically been spearheaded by designers, sometimes artists, and, to a lesser degree, celebrities. Among these, it was the designers who ruled the trend universe.

TEACHING VERSUS LISTENING TO THE CUSTOMER: BUILDING THE FOUNDATIONS OF MANIPULATION

The fashion industry is based on a lie. It is this lie that drives its business models. The lie is that customers demand fashion, consequently implying that fashion is a consumer-driven industry in which fickle customer tastes change rapidly and garment producers must respond quickly to such changes. The lie is promulgated by industry specialists, scholars, analysts, and enthusiasts who have accepted the mantra that the change in customer tastes commands fashion producers to respond to changing tastes and preferences. Hayes and Jones (2006), using summaries of prior fashion trend research, conclude that the definition of fashion denotes elements of style that are short-lived. The authors posit that they

[handwritten margin note: companies through the media make societal pressures]

[handwritten margin notes: "How do we know this? Do the retailers who are manipulating the media claim this?? → BIAS"]

are short-lived because customer tastes change quickly. It is, supposedly, this erratic customer propensity to look for constant change to which the industry must respond. Other researchers who study the historic developments of fashion style, notably Christopher et al. (2004) and Richardson (1996), explain that the core feature in defining what is fashion and what is not is found in the definition of what consumers understand as style. Style is different from fashion because what is stylish may not be fashionable at the moment. It is a time-sensitive convergence of a critical mass in market understanding and agreement of what is stylish and current that makes an item fashionable. It is all based on constantly changing tastes. But it is not the consumers who first change their minds on what is fashionable. It is the fashion trendsetters who drive the change cycle, and these fashion trendsetters are, and have always been, industry insiders. The fashion industry, not its customers, dictates the change of trends. Therefore, fashion sales are *retailer driven*, not *consumer driven*.

[handwritten margin note: "How do influencers play a role in this?"]

However, we are led to believe that the change in fashion tastes is external to the industry. It is, supposedly, based on esthetics. Cultural forces such as art, music, and even politics shape people's esthetics. Based on that logic, the expectation is that when a cultural change occurs, a new trend is born. As the trend gains popularity, fashion customers demand its reflection. But the truth is that customers are told what to demand. In addition, they are also told how to consume.* Fashion trends are not born from cultural factors, but are strategically created in order to fit a very precise eight-season production calendar. Although today there are many more than eight fashion seasons, the structure of eight main time frames remains valid in terms of trendsetting. Davis (1992, p. 28) puts it simply: "fashion is the continual, regular, and institutionalized change in dress and adornment." The key point is "institutionalized."

[handwritten margin note: "Could it be different now?"]

Change in trends drives industrial growth, but the change in trends is not because of customer demand. It is because of retailer dictates. Therefore, the fashion industry is a retailer-driven industry, not a consumer-driven industry. As famously stated in the 1992 BBC documentary *The Look: Power of the Press*, "Nobody knows what's nice. Or not … It is all a matter of opinion." This statement encompasses the truth that customers have their tastes dictated to them. The six-part documentary series is set in the heyday of the "super models," the "super brands," and the "iconic"

[handwritten margin note: "influenced by mass media"]

---

* The institutionalized teaching of fashion consumption patterns is the core subject of Chapters 6 and 7.

fashion publications. The early 1990s were the years of the fastest proliferation of fashion globalization. The documentary explores why, by that time, the fashion industry had become a main international cultural and societal focal point.

A few factors were at play, such as technological advances in telecommunication, globalization of markets and culture, and the goal to sell products based on successfully exporting wealthy, mostly American, popular culture images through the high-quality visual power of satellite, cable, and other new media. Media were changing, and with them, customer impressionability.

Through the diversifying and growing media channels, trends were skillfully and strategically promoted at increasing rates, but with decreasing life spans. Trends spread globally in months and remained relevant for a year. Some elements of a trend could last a bit longer, but not more than two calendar years, after which something new was needed. This new trend was needed in order to make additional sales.

Additional sales drive profitability in all industries. In fashion and its various complementary sectors, novelty not only has to be present; it also has to be sold. It is sold through the blessing of knowledgeable critics whose taste authority deems a creation innovative, forward, fresh, and, most importantly, beautiful. Or not.

The *Power of the Press* BBC episode that focused on fashion critics and their press releases explained the symbiotic relationship that had developed between fashion writers and designers—the trend dictators and creators. The main point was the tacit understanding that both parties exist in an ongoing career partnership in which discourse is discouraged. Therefore, fashion journalists had an incentive always to find something nice to say about a fashion collection. The success of a collection depended on two paragraphs of carefully worded copy. Billions of dollars depended on a sound bite. In this chain, a perpetual trend of adulation was created, in which fashion journalists continually liked almost everything the main fashion designers put out each season, thereby cementing the power of the few leading fashion houses as the trendsetters of the world. The market ended up demanding whatever they deemed fashionable during a season.

This culture created astounding statistics—a trillion-dollar industry touching every human being on the planet was driven by a thousand or so people who got together twice a year for 3 weeks at a time to view a few collections and determine what they all wanted to sell. During September/October and February/March, those one to two thousand humans would be cramped into tents, sit on plastic chairs, and decide what the upcoming season's trends would be.

Then and now, not just anyone can get into the tents. Attendees are carefully selected. The shows are by invitation only, and the design houses issue the invitations. They are sent only to decision makers who can create sale increases. Limited space dictates that only the most important industry representatives can be present. Among them are not just trendsetters, but also the manufacturers—from fabric and textile conglomerate owners, to button, zipper, and toggle factory owners, to perfumers, leather manufacturers, and, of course, publishing houses. After all, it is the pages of fashion publications that have always dictated trends.

It is undeniable that throughout fashion's history, the power of fashion magazines has been the pillar on which fashion promotion rests. It is ironically said in industry circles that the difference between fashion and clothes lies in what each industry sells. The tongue-in-cheek answer is that the clothing industry sells clothes, while the fashion industry sells anything but clothes: mainly, magazines.

Fashion magazines sell fashion trends through carefully constructed images (Phillips and McQuarrie, 2011). These constructs come in two main formats—fashion editorials and printed ads. Fashion editorials are the "story in pictures" of a trend put in a cultural context. Editorials focus trend formation. They put the clothes of selected brands in context, and explain through environmental settings such as special sets, landscapes, cityscapes, and even political venues how the garments relate to current cultural realities. Printed ads directly showcase the individual fashion brands. Their main point is to stress the name value of a brand, its image, and its importance. Trends, although visible in printed ads, are sometimes hard to define and more prone to individual interpretation because printed ads are often surreal and indirect. Their goal is to catch the reader's attention and relate it to a name, not a changeable trend that is replaced often. Printed ads are there to create brand awareness, not trend awareness.

As the globalization of trends escalated in the late 1980s and 1990s and the world moved toward the age of technology, through the pages of fashion magazines, trends diffused internationally in decreasing time frames. The magnified media channels enthusiastically promoted each new trend, leaving society to respond to the glitz, color, fantasy, and glamor of it all, and allowing fashion to become more than just an industry. It became one of the main pillars of the new global culture. The sale of these fashion magazines also became global. The fashion press became the core venue of the globalization of trends. Trends were carefully synchronized, and that synchronization was used to promote them systemically around the world. This collusion was driven by the commercialization of page space. Appearing in the pages of the leading fashion magazines became the defining currency of commercial success. Where exactly a brand appeared on those pages signaled its importance in the industry.

Fashion magazines raise revenue in two ways. One is the direct sale of printed copies, which for most magazines is either a break-even or a losing revenue stream nowadays. It is less profitable than it was because it has been greatly impacted by the advent of the Internet. The other way fashion magazines make money is through selling advertising space in their pages. Just prior to the proliferation of the Internet and its gradual improvements in the delivery of high-quality imagery, fashion magazine sales had escalated from the late 1980s to the late 1990s. Appearing in their pages was (and still is) of the utmost importance for visibility of fashion designers, accessory manufacturers, retail chains, and, of course, the models whose images sold both the items advertised in the pages of the fashion magazines, and the pages themselves. That visibility was priced accordingly. Depending on placement, either a full page of direct advertisement or a page to be included in fashion editorial spreads could sell for $10,000–$50,000. The prices varied, based on the prestige of different publications.

This revenue was mostly profit to the publishing house, making the commerce of fashion print at least as vital to the industry as the sale of clothes. Both industries increasingly internationalized as markets liberalized. The growth of both depended on the imagination and creativity of a third main industrial player—the advertising agencies (Aspers, 2012). Leslie (1995) tracks the internationalization of the advertising industry and explains how the biggest agencies, primarily headquartered in the United States, Japan, and Great Britain, became increasingly reliant on their international revenue as their affiliates expanded globally. As a result, their campaigns fueled the "internationalization of consumption" (Leslie, 1995, p. 402).

As the 1990s amped up, the creativity and ability to seduce consumers with the power of image needed to be just as innovative and rapidly changing as the clothes and magazine editorials themselves. Neither clothes nor magazines would sell without the visual connection that elicited an emotional impulse to covet, dream, fantasize, escape, aspire, and, eventually, purchase the magazines. The pages of the magazines showed the promised path of actualization by dictating the possession of the right clothes, accessories, and cosmetics (Leiss, 2013).

A collusion of agreement is needed for this industrial "machine" to run smoothly. Sales increase when promotion is done in tandem. The technical term is brand synergy. It describes the symbiotic codependence of all items in a fashionable look. Such items are, of course, the clothes the model is wearing, but also the cosmetics, the shoes, the accessories, and the jewelry. Additionally, the props need to be synergized in the style of the "look" and "feel" of the image. Therefore, the headphones, or laptop, or bike, or the hotel in front of which the model is standing must not only match, but also serve a promotional purpose. All items benefit from the promotion; therefore, all colors must match, the right brands must be showcased, and the right amount of balance between item and background must be achieved. Sales around the world depend on the understanding that the right balance of blues and greens in an image will be changed to the right balance of reds and yellows in the image that will be created for next month's magazine issue.

The interconnectedness among products with respect to marketing became so prolific that industry professionals coined the term *brand alliance* to describe how fashion and related products are promoted as a whole. Brand alliance explains the process of managerial product positioning in promotional platforms that showcase several brands as a part of the same image (Gammoh et al., 2010). In fashion, a whole look is promoted, as opposed to individual items. Then, a multitude of looks are promoted as a lifestyle. For example, in a typical print campaign focusing on a garment or an accessory (such as a handbag), more information can be found in finely printed bullets or footnotes about the cosmetics and makeup of the model than about the actual item that is the focus of the campaign. In the creation of the ad, marketing teams representing all individual items are involved to ensure complementarity in promotion. This complementarity became essential as brands became global.

GLOBAL BRAND PROLIFERATION

Knight and Cavusgil (2004) state that global brands may be the most readily observable outcome of globalization. Global brands do not just appear;

they evolve over time. In fashion and related industries,* the mega-brands are well known. Some, such as Chanel, Gucci, Yves Saint Laurent, Givenchy, and Burberry, are fashion legacies, well established prior to the proliferation of modern globalization. They were able to embrace globalization's advent and use its vehicles to maintain and increase international market leadership (Haig, 2006). Others famous brands of the past, however, such as Fiorucci and Holston, today are merely names to be remembered by fashion historians.

Among the most famous brands today are companies that were born because of the globalization of imagery, fashion, and culture. Scholars use Armani as the leading example of the creation and proliferation of a global brand. Both the man and the brand escalated in popularity after providing Richard Gere's wardrobe in the 1980 film *American Gigolo* (Okonkwo, 2007; Tungate, 2012). From then on, the brand grew and diversified, and remains a leading retail powerhouse, selling different clothing, accessories, and home product lines.

Another example is the house of Versace. Its founder, the late Gianni Versace, is hailed as the master of global promotion. Among the many innovative tactics in international advertising that Versace brought to the front of fashion marketing is the promotion of the faces of the trends. Versace is credited with inventing the supermodels by understanding that promoting their images internationally allowed his brand to rise to global leadership (Wilcox et al., 2002).

Versace remained virtually unknown from the founding of his company in 1978 until the late 1980s, when his unique promotional tactics catapulted the brand straight into global stardom, bypassing the establishing of market leadership in his home country of Italy. If anything, established Italian and European fashion houses denigrated his look, calling it garish, tasteless, and cheap. But the image of multiculturalism portrayed in the multiracial and multiethnic models Armani promoted by highlighting their nonmainstream look spoke to global audiences, and the brand's popularity exploded. Versace also focused his campaigns on the glamor of the industry and the theatrics of it all. His shows became spectacles beyond the runway, with celebrity-studded after-parties in glamorous settings, where photographers documented the entire scene and, of course,

---

* As noted, the terminology "fashion and related industries" is used to include apparel, cosmetics, fragrance, shoes, and accessories—all are interdependent and, it is argued here, inseparable in promotion.

the clothes. Versace used these images as press promotion, as vital to the marketing of the brand as the fashion shows, magazine editorials, and print ads. Versace changed fashion promotion because the house chose to promote the industry's lifestyle. The general public embraced the message and the celebrity of fashion designers, models, editors, and stylists.

In the 1990s, on the heels of the new powerhouses of Armani and Versace, other superstar designers such as Todd Oldham and Isaac Mizrahi appeared on the tightly guarded fashion scene. However, they eventually disappeared from the forefront of the global fashion spotlight. Like Fiorucci and Holston before them, their fame was short-lived, and their innovative and creative impact was absorbed by the survivors. The impact of these designers, though much celebrated and lauded on their arrival, diminished because they could not sustain global profitability. Globalization increased exposure; it increased the incentives for newness; but it also increased the importance of money. International promotion, production, and distribution are much more expensive than localized commerce.

Going global escalates costs and debt. Low and Mohr (2000) show the importance of advertising, in particular during the global expansion of a brand, by looking separately at advertising budget allocations relative to sales promotion for a multitude of branded firms. Their findings indicate that higher advertising budgets lead to more favorable consumer attitudes and build brand equity. The importance of advertising was no news in the dynamic 1980s and 1990s, but the price tag of advertising globally to ever-increasing market segments escalated, and only the most profitable fashion houses could afford to keep increasing their advertising budgets, crowding out any newcomers and establishing new partnerships. The proliferation of media other than fashion magazines helped in the advent of such new partnerships and led to the growing importance of integrated marketing.

As the Armani example shows, the power of film was well documented in creating and promoting a global brand. From the 1990s onward, the power of television and other media in addition to film increased, with the growth of cable programming, international syndication, and the advent of social media. Fashion-brand promotion was experiencing a massive change. It moved beyond the pages of fashion magazines into multimedia product placement platforms. Product placement is putting a brand (usually several products of the same brand) in actual life situations as exemplified by nonprint media, such as television, film, and social media, and

[handwritten marginalia: Someone loves Stranger thing → want the sweatshd → consumer → trend]

social events, such as sporting, music, and even political venues. This new reality of integrated marketing increased the ability to induce consumers to form an emotional attachment to products because, particularly in fashion, the products were not just parts of a printed fantasy; they became part of real, palpable situations.

Saladino (2008) tracks those changes and concludes that product placement had moved from an "additional way to market" to a key aspect of promotional strategy, more than tripling its monetary impact from $2.2 billion in 2005 to over $7 billion by 2010. Among the examples of success in product placement strategy is the impact the popular HBO series *Sex and the City* had on the brand Burberry (Blackmon, 2007; Phan et al., 2011). The image of the main heroine in a Burberry coat during an emotional scene did more for the brand, in terms of cementing its relevance to style, than all the firm's prior advertising efforts. In the dynamic social media age of the 2000s, Burberry had relentlessly tried to revitalize a stagnant and (the worst possible description a fashion brand can bear) old-fashioned look. That one image changed the relevance of the ubiquitous Burberry earth-tone plaid from old-fashioned to classic in a TV instant. This example shows the quiet power change that had occurred in the industry with the advent of new media platforms.

Although television is not necessarily a new medium, cable television is still considered a communication platform that is not mass-market contingent because it is fee based. However, components of its programming can be proliferated today with incredible speed and scope. Images from programming in terms of styling and setting can easily be shared on social media. In this way, their fashion looks can proliferate further than the viewership of a particular network or program. This proliferation elicits social media discussion, and in those discussion forums, fashion enthusiasts offer opinions, discuss features, share experiences, and connect to the average fashion buyer in ways that the old trendsetters could not. The new trendsetters are immediately connected to the customers. The old trendsetters had to rely on forecasting firms that relentlessly analyzed consumer behavior and satisfaction to offer information on what customers demanded.

Today, nonfashion magazine "fashion" media offers a voice to new trendsetters as well as old trendsetters. The new trendsetters are the stylists of film, television, theater, and music performers. The new trendsetters communicate their trend analyses mainly through social media. As the 2000s amped up, mobile communication technology platforms increased the immediacy of fashion information exchange. No longer subject to carefully placed airtime advertising slots on television, or even more rigidly delivered (and very expensive to produce) monthly print campaigns, the delivery of a fashion statement became increasingly instantaneous. Furthermore, anyone could make a fashion statement and deliver it successfully. In this reality, stylists took on a new and decisive role in trendsetting.

Through savvy use of information technology, stylists rely on celebrities rather than fashion designers as the new trendsetters. Stylists dictate trends based on happenings in lifestyle hot spots, clubs, and fashion *flash points* (Barn and Lea-Greenway, 2006, p. 261). They follow the designers, but also offer their own voice on how to pair and promote designer looks. They are essential pillars of integrated marketing, because they promote several brands in tandem. The term describing this promotion is *brand adjacency*, referring to how individual pieces of apparel from different brands can best work together in creating a trendy look. This multibranding embrace gives stylist credibility.

Stylists base their influence precisely on the fact that they are not tied to any one fashion house or major brand label. Stylists position themselves as hip and progressive trendsetters, more attuned to shoppers' needs, and more committed to bringing the customer value and utility. The promotion of their impact is on newness. It is also on unpredictability. No longer bound by the calendar of fashion week, stylists can promote a flash point look throughout the year, without waiting for the (ubiquitous in the past) *Vogue* September issue to tell them the fashion future. Retailers adjusted accordingly, competing to respond to market demands created by stylists outside the rigid fashion calendar. Fast-fashion trendsetting was born.

FAST-FASHION RETAIL

Fast-fashion trendsetting became possible due to easing of the time constraints borne by traditional fashion merchandising. These time constraints were defined by the life span of trends, and their lengths depended on how long industry specialists thought a garment needed to remain on store shelves before it became outdated. As a garment became less trendy, it went "on sale." The "on sale" model of retailing has been the foundation of decades of fashion

TABLE 1.1

Traditional Pricing Sheet, US Mass-Market and Mid-Market Garment Retail

Production Link	Retail Price (%)
Fabric	13
CM (labor, overhead, profit)	6
Trim	2
FOB	21
Duty	4
Clearance and inland freight	1.4
Import office costs	6
Total production and transportation costs	53
Retail markup	100–53 = **47**
Retail markdown	**30**
Net retail sale	**70**
Net profit	**18**

Source: Based on Birnbaum, D., *Crisis in the 21st Century Garment Industry and Breakthrough Unified Strategy*. The Fashion Index Inc., New York, 2008.

Note: CM, cut and make; FOB, freight on board. The bold numbers indicate the percent in the pricing of the garment that is the profit margin for the retailer.

commerce.* Then, fast-fashion retailers revolutionized the industry when they took the concept of "sale" out of selling. "On sale" became "no sale."

"On sale" had been the traditional trend barometer. When items were trendy and in demand, they were priced high. As the trends waned and merchandise became stuck on shelves, "on sale" signs appeared to entice additional speedy sales. The "on sale" culture defined retail pricing strategies. Table 1.1 shows how traditional apparel pricing worked prior to the revolutionary approach of the "no sale" tactic.

Profitability depended on the markup/markdown balance. Merchandising success depended on the ability of a retailer to make many sales before a trend waned during the markup stage of sales. Once the trend began to set, the markdown game began, in which garments went on sale, the percentage gradually increasing from 10% off to 50% off when items went on clearance. As Birnbaum (2008) calculates, the average markdown equalizes to about 30%.

This business model developed when trends had relatively long life cycles. A look, or an aspect of a look, such as pattern and cut, had an

---

* The legacy of the "on sale" model, from its advent to being embraced as the most profitable way to increase sales, is the focus of Chapter 6.

average lifespan of 2 years up until the late 1990s. It took approximately 6 months to diffuse globally. In the eight-season, tightly scheduled fashion calendar, garments could be manufactured based on similar look characteristics for up to one full calendar year. For example, even though jeans were manufactured and stocked for both the main summer and winter seasons, they could remain flared, bootcut, or of any other design that was deemed fashionable at the time. Therefore, during the production process, item differentiation could remain minimal. During the markdown process, as new merchandise was brought out, older items that remained branded and had the main trend elements would also sell, spurred by the better value the "on sale" sign promised. Customers gained validation in addition to value.

The validation came from the fact that the new items would follow a similar look, already set by the trend, and cost-conscious consumers would be more apt to purchase an on-sale item if they were convinced that the item would remain fashionable. Research on strategic fashion retail examines this dynamic. An example comes from Robinson and Doss (2011), who survey consumers on their evaluation of fashion prestige substitutes. These are the substitute options offered for the most prestigious item available. The authors explain that the process is perceived by consumers to carry a high risk. The risk stems from the transaction costs of substitute evaluations. Therefore, in retail, a signaling strategy based on constant communication of value reassurance is essential for the efficient moving of merchandise off the shelves. This value reassurance had become such an important sales tactic as to lead to the fact that the bulk of all apparel sales occurred (and for non-fast-fashion retailers continues to occur) during the markdown stage (Cachon and Swinney, 2011). Traditionally, the markdown process took over a full calendar year—enough time to entice additional sales by the gradual, albeit unwilling, lowering of prices.

In any industry, the buying behavior of customers dictates sales. In fashion retail, that behavior has gradually become habitual. The quest to understand the buying habits of shoppers is behind the collusion in trend dictation. Knowing that only a finite number of sales can be executed profitably necessitates that trend changes be carefully managed. That is why, during the time of traditional fashion retail, only a few fashion design houses and their fashion journalist and editor friends ruled the industry.

During the decades of traditional (as opposed to fast) fashion retail, fewer and carefully timed trend changes were easy to manage in production, so that final sales would be most efficient with respect to risk of financial loss. If a trend suddenly appeared, and merchandise that had been carefully designed, promoted, and manufactured in large volumes was deemed unfashionable, the markdown process would lead to lowering prices faster than desired, and increasing volumes of merchandise would have to be sold at a loss. Therefore, the incentive for changing trends was one of predictability, not innovation.

The hypocrisy of it all seems laughable, in light of the public image of the industry that promoted constant change, an innovative approach to esthetics, and a strong commitment to responsiveness to customer preferences. The pretense was that the industry was looking for change because sales were driven by constant change. The truth was that the change needed to be marginal, not too innovative, and conveniently managed. The business model of "on sale because we can" versus "on sale because we have to" made branded retailers such as Macy's, Lord and Taylor, and Sears the leaders in the conglomerate corporate structure of globalized merchandising—until Hennes & Mauritz (H&M), a small, insignificant, and decidedly uncool at the time Swedish retailer revolutionized the industry (Giertz-Mårtenson, 2012).

Embracing the new platforms of direct marketing that emerged and evolved in the 1980s and 1990s, H&M started its successful expansion by introducing a new way of retailing—no "on sale" signs anywhere to be found. Instead, clothes were promoted directly as fairly and valuably priced, and in addition, they were not just clothes, but fashionable items advertised by famous models. H&M started to sell online earlier than most, in 1998, and as the company focused its efforts on reaching customers through direct channels, it was obvious that the "on sale" tactic was not going to be successful online because there it lacked context.

"On sale" is enticing in traditional retail environments because customers can compare the quality, texture, and fit of the garment next to options that are not on sale. When the similarities of quality are significant, and at the same time, the price discount is significant, then shoppers are more apt to succumb to their buying impulses. They justify the value of the on-sale item through the amount saved in relation to all its beneficial attributes. However, in the online environment, such spontaneous benefit–cost analyses are not present, because the temptation a higher-priced item offers is not tactile. Even though online retailers can position items at different price points, the promoted latest and most fashionable garment is not next to a cheaper version. It is a few clicks away, placed in a different context. Furthermore, online shopping carries a level of risk because of uncertainty of fit, which lowers buying impulses. Online shoppers are also more apt to change their mind against a possible purchase during the decidedly unpleasant ordering process, in light of high shipping rates and delayed gratification, since they have to wait for the garment to arrive. For these reasons, online shoppers could be enticed most effectively through one main tactic—low prices above all else. H&M's model was revolutionary because it decided not to waste time and money during the markdown process, but to remove the markup/markdown stage completely from garment pricing, thus pricing clothes at much lower points, closer to production and transportation costs.

Critics were adamant that the approach would be ultimately unsuccessful because of a retail tradition that was based on impulse purchasing (Bayley & Nancarrow, 1998). But others soon followed, such as Zara and New Look, representing what Ünay and Zehir (2012) describe as the new wave of innovative fashion entrepreneurship.

Spanish conglomerate Zara, part of Inditex SA, has been the model of fast-fashion retail efficiency, according to industry supply chain and manufacturing scholars (Lloyd and Luk, 2010; Tokatli, 2007; Tokatli et al., 2008). In a *Wall Street Journal* article, Rohwedder and Johnson (2008) explain that Zara's success is based on superior supply chain management that allows it to lure wealthy customers away from luxury brands. It is able to provide the latest trends in fashion-forward merchandise that looks expensive in record time, and this ability lies in its fully integrated model, in which all aspect of production are managed with a focus on speed. This full integration is driven by its parent company, Inditex, currently the biggest fashion group in the world. It owns its textile manufacturers and other backward links (as they are called) in the production chain, and through exclusive partnerships and licensing arrangements, it offers complete fast-fashion differentiated product lines in clothes, accessories, and home products through managing such brands as Massimo Dutti, Stradivarius, Pull&Bear, Bershka, Oysho, and Zara Home.* Massimo Dutti specializes in more stylish designs simulating high-fashion choices; Stradivarius is lower priced and focused on young styles; Pull&Bear has an urban look and provides more male options; Bershka offers boys and girls casual activewear; Oysho offers undergarments—fast fashion's response to Victoria's Secret—and Zara Home is the company's version of Ikea. This level of diversification allows Zara's parent company Inditex to follow the

---

* For a detailed breakdown of holdings, see Rohwedder et al. (2008).

integrated, multibrand-promoting legacy of luxury conglomerates such as Luis Vuitton and Christian Dior's parent company LVMH, formerly Moet-Hennessy Louis Vuitton.

Luxury conglomerate multibranding was done with the goal of developing product breadth across a lifestyle of luxury where dressing, accessorizing, bathing, grooming, decorating, eating, and drinking were interlinked in one company that offered all options for a dream lifestyle. By definition, luxury multibrand proliferation is priced high. Therefore, luxury sales are elastic and prone to recessions and market uncertainty (Ait-Sahalia et al., 2004). They are also easily tapped out in demographic terms. There are only so many buyers in a market who can afford luxury items. Growth is contingent on convincing the same pool of buyers to make new purchases. Unless rapid economic growth moves sizable portions of the population into income brackets of considerable wealth, luxury markets do not expand quickly; or, at least, they did not traditionally, and only a few conglomerates managed to corner the global fashion luxury lifestyle market.

Globalization helped those few brands to proliferate fast. For example, Louis Vuitton had two stores in 1970. Today, it has over 400 stores worldwide. However, the fast-fashion conglomerates dwarf that number. Tokatli (2007) estimates that by the mid-2000s, Zara, H&M, and Gap had around 1000, 1400, and 3000 stores, respectively. These estimates are conservative, because much more growth in fast-fashion store proliferation has been reported since (Joy et al., 2012). Rohwedder and Johnson (2008) report that Inditex's total number of stores in 2008 was 3691 in 68 countries. In 2012, the *New York Times Magazine* estimated that the number had grown to 5900 stores in 85 countries, and explained that the number is always changing because Inditex opens more than a store a day, or about 500 stores a year (Hansen, 2012). According to the article, Inditex opened 400 stores in China in 2012. Discussing this aspect of growth in new retail outlets in the context of the economic boom in South East Asia, the BBC journalist Alex Riley narrated the three-part documentary *Secrets of the Superbrands*, which aired in 2011. In the episode devoted to clothing *superbrands*, as they are dubbed in the film, Riley says that Adidas is on schedule to open two new stores a day in China for "the next 5 years."*

As wealth is growing today, particularly in the developing world, sales there are growing for both luxury and fast-fashion brands. However, although traditionally fashion and related good conglomerates enjoyed

---

* For information on the documentary, go to: http://www.bbc.co.uk/programmes/b011fjb5.

a certain level of market monopoly, today the fast-fashion retailers are increasing the competition. Albeit they are creating their own monopolistic structures through thoughtful vertical integration, they are successfully luring customers away from luxury brands. Therefore, in terms of value, luxury conglomerates have fallen behind fast-fashion powerhouses such as Inditex. Inditex, H&M, New Look, Mango, Forever 21, and others offer a whole lifestyle as well as much lower prices. They promote a "cool" lifestyle: not a "luxury" lifestyle, but a lifestyle that offers all the staples of belonging and social acclaim that luxury clients showcased. It is now more important for modern shoppers to indicate they are hip and progressive, rather than rich.

2

The Promotion of a Lifestyle

ADVERTISING BEHAVIOR, NOT PRODUCT

For decades, fashion promoters have gradually focused more on advertising whole looks. As a result, people integrate products promoted in such a way in their daily consumption habits. In that way, it can be said that we all "live" fashion, not just wear it. Interrelated consumption of products that have style as a main value-adding feature depends on updating that style (Tungate, 2012). The frequency of updates has escalated as the life cycle of trends has shrunk, and fast fashion has introduced a whole new calendar schedule by which modern-day shoppers update their personal style.

Fast fashion has become a social phenomenon fully integrated into daily behavioral patterns, because while a person may not purchase clothes daily or weekly, a person purchases fashion-related products semi-daily (Gabrielli et al., 2013). Female shoppers in particular purchase cosmetics, fragrance, accessories, home products, and even electronics and other fast-moving consumer goods (FMCGs) with increasing regularity, leading to the high volume of FMCG sales. Most, if not all, such FMCG products are promoted for their style, look, and trendiness. Their advertising is presented in relation to fashion images that are integrated in terms of clothing, accessories, health and wellness maintenance, and décor. This is the case because fashion promotion advertises lifestyles, not only pieces of apparel. Therefore, the impact of fashion promotion influences the daily consumption decision-making of ever-increasing marginal consumption. Marginal consumption is the concept of the additional purchases shoppers make on a regular basis.

Brand identity drives this buying behavior. Today, brands have evolved to represent more than the makers of goods (and in certain cases services) with respect to quality and trust (Carroll, 2009). Brands have traditionally added a feeling of belonging to an identity that was external to individual consumers. Their uniformity removed incentives for individualization of expression in image, because the image was dictated to consumers, who had to embrace the look of a brand.

In the new reality of fast fashion, Carroll (2009) argues that brands do exactly the opposite. The increased number of options fast-fashion brands provide creates incentives for consumers to engage in unique self-expression and build their own self-identities. Consumers have more options because they can afford to make more purchases. In addition, manufacturers today provide an increasing number of choices by constantly and consistently differentiating fashion and related product lines.

At the core of the ability to offer this level of differentiation are low prices. The ability to offer an increasing variety of options lies in their successful commercialization. In other words, manufacturers offer many choices when they can make profits in the sales of those choices. For consumers, choices not only provide style options, but also lead to lower prices, because increasing choice increases the competition for sales among retailers. The end result is gradually decreasing prices. The outcome is a retail environment in which the competition among producers is based on price. All offer a high degree of choice, because modern technology allows speedy duplication of ideas and the efficient replication of those ideas in commercial products. In effect, all fashion merchants today, whether they follow a fast-fashion or another traditional retail model, provide consumers with product lines defined by constant updates and change. These updates create an unprecedented level of choice, and the presence of choice changes the competitive platforms of the industry. The main change is that the competition is not based on unique design, or other style attributes that can entice consumers under the assumption that a specific retailer can offer the most fashion-forward choice. Since choice is no longer a competitive attribute, the only way retailers can engage in competition with each other is to try to beat the low prices of their opponents. In this competitive environment, it appears that customers reap major benefits, as merchandise is increasingly affordable and varied.

HaeJung (2012) posits that the ability of modern-day shoppers to exercise higher degrees of choice through their fashion purchases has a hierarchical structure in terms of a hierarchy of needs. The author argues that today brands contribute to self-actualization, which has transitioned the fashion-brand experience toward one unified global archetype of consumer experience. The argument is that the social platforms of belonging are expanding culturally, and individuals in different countries exhibit patterns in their commercial behavior that reveal their brand preferences,* which links them in a similar fashion, so to speak. These social platforms of belonging, as expressed through the way people dress, are expanding and at the same time converging into a global behavioral entity. This outcome is due to the globalization of commercial branded sales.

Brand proliferation is the industrial term that describes the global expansion of brands and the augmentation of different product lines that bear the same brand. That augmentation today transcends borders and provides a uniquely converging platform of consumption options. Although there are increasing numbers of products, they are classified in a standard format through their branding. As a result, internationally, consumers are aware of certain labels, firms, and products that stand out to all people in all nations. This awareness builds connectivity in developing tastes and preferences. This development process happens through the change in trends.

As trends are introduced, absorbed, and abandoned in the modern world, customers may have a more directional role in that process of creative destruction. The term *creative destruction* has become popular in describing commercial activity that is based on high degrees of innovation. It is applicable in fashion because, although it may be contentious to posit that style trends are really innovative, they are changed often. As a result, new products are brought out to store shelves and promoted for their novelty. They are frequently replaced, and in the process, information on their popularity is analyzed. In this information analysis, customers feel a level of creative equality because their own point of view can be reflected in trendsetting with greater immediacy. All of this is possible because trendsetting information is no longer unidirectional, but increasingly relies on user-generated content (Kulmala et al., 2013).

User-generated content growth is contingent on fairly recent advances in information technology (IT). Today, social media and all the devices deployed in its use provide platforms for information exchange among consumers, among producers, and among promoters of fashion that allow increasing numbers of individuals to express their style points of view. It is important to remember that this reality is just over a decade old and

---

* The expression "commercial behavior" with respect to brand preferences includes direct purchases, which are the measurable outcome of strong brand allegiances, but also such concepts as brand awareness, loyalty, or lack thereof.

for many even younger than that, as it all relies on the World Wide Web. Not until well into the early 2000s was Internet access considered widely available, and even then this only held true for people in the developed world. This fact is a main topic in the literature on the global digital divide, as the reality of low IT infrastructure in poor regions is described (Chen and Wellman, 2004; Guillén and Suárez, 2005). Digital divide issues were, and for some still are, pervasive even in developed nations (mainly in the United States) where significant portions of the population live in rural areas far from major metropolitan centers. These problems stem from infrastructure underinvestment in connectivity tools such as fiber-optic cable. However, digital divide issues gradually subsided in the academic discussion sphere with the advent and rapid performance improvement of smartphone technology (Qureshi, 2012). Today, the Web is easily brought to even the most remote of locales through smartphones, because it is much easier to construct one cell tower than to lay miles of power lines and fiber-optic cable (Ali, 2011; Techatassanasoontorn and Kauffman, 2005). The fact that from the early 2000s until today a whole body of literature on digital divide issues, examined across disciplines,* grew and then subsided in magnitude is a testament to how fast technology changes our reality. It is almost unimaginable to see the television commercial for BMW aired in the United States in early 2015, during which, in a clip dated 1994, famous American TV journalists Bryant Gumbel and Katie Couric discuss this new invention called the Internet. The example is important to offer in relation to fashion marketing, because during that time—the early 1990s—the fashion global megabrands were at the height of their international proliferation. Their power was vested in their role as the dictators of trends.

Today, the new technological platforms of communication have allowed new voices in trend creation to emerge, be heard, and make an impact. It all sounds very progressive and liberating, and when it comes to creativity alone, that is indeed the case. In a way, today all individuals touched by fashion have an unparalleled ability to influence trendsetting. Input is immediate and in many cases emboldened by the relative anonymity of social media. For these reasons, virtual input into fashion has become the core foundation of modern user-generated content. User-generated content is best illustrated by fashion blogs.

---

* Most notably, sociology, international business, information technology, and international marketing provide the core body of digital divide research.

Fashion blogs allow consumers to provide major critical content through comments, discussions, and story sharing. The particular subject of discussion is set forth by the blog creators. Depending on the frequency of blog entry activity, the subject of input changes regularly and is usually broken into fashion categories, accessory categories, fragrance, cosmetics, and even lifestyle input on home product choices. For example, a fashion blog's weekly discussion could be all about hats. The following week, it could be all about coats. The next week, it could be all about "the" scents and lotions of the season.

Fashion blog creators are stylists and other industry enthusiasts, and their following depends on their own image as fashion savants. As such, they dictate styles and promote brands. Consequently, the fashion industry has shown great interest in fashion blog marketing (Kulmala et al., 2013). The interest is commercial in two ways. One is that bloggers can become spokespersons for particular brands without appearing as such. This fact adds credibility to their promotion, because it does not seem as though the fashion labels are paying the bloggers for the promotion, but rather, that the bloggers impartially discuss brand attributes based on utility. The second way fashion houses commercialize the blogging universe is through direct advertising by purchasing advertising space on individual blogs.

Additionally, fashion blogs offer the immediacy of mobility. The proliferation of mobile communication technology has fueled their rise in trendsetting importance. Magrath and McCormick (2013a) note that 1.7 billion smartphone users are estimated to be behind the close to $1 billion sales increase for fashion products through "encouraged consumer spending" resulting from the use of smartphone apps. Continuing their research on smart media's impact in fashion promotion, Magrath and McCormick (2013b) add that mobile fashion commerce has tremendous potential for growth because it represents fewer than 5% of online sales, which in 2011 were estimated to be around $120 billion. This number is impressive because, as already noted, total world consumption of clothes is estimated at $1.3 trillion. This growth is contingent on brand awareness that is global. The numbers are so impressive because fashion products are sold across countries, and consumer information on their market performance is not confined by the old-fashioned country-specific limitations of national markets. Global branding is the defining feature of modern-day fashion economics and, much like IT's legacy, its advent is fairly recent.

GLOBAL BRANDING

As early as the late 1980s, global brands began to be discussed as the future of advertising, and were lauded as a great resource savior because they would eliminate the need for costly segment advertising (Domzal and Kernan, 1993). Segment advertising had been the promotion platform since the advent of mass production across industries. It targeted specific markets, well separated by geography, ethnicity, and culture. But globalization and technological innovation started to blur such separation. Culture was becoming global.

After The Beatles appeared on The Ed Sullivan Show, elements of cultural diffusion started to grow, escalating with the improvements in satellite communications and the proliferation of the American entertainment industry internationally. By the 1990s, it was said that the biggest American export was culture (Craig and King, 2002; Kraidy, 2005). From movies to television programs to music and Internet website platforms, America is singularly the main global culture that permeates every other country's national culture (Crothers, 2012; Pells, 2011).

American films, television shows, and magazines (including fashion magazines) promoted a lifestyle that was very different from that in most other countries, including industrialized Western nations. As more channels of mass communication appeared and improved, such as cable television and the resulting diversification of specialized programming, more and more people around the world were exposed to a lifestyle that demonstrated a higher quality of life than their home environment. Homes were bigger and full of the latest tech devices, everyone (including teenagers) drove the latest cars, all children had their own rooms, all young women were beautiful and dressed, accessorized, and behaved inspirationally—freer, happier, emancipated (according to international standards at the time, but it can be debated whether they were really emancipated or just objectified in a different way).

In cultural terms, this dream lifestyle spoke volumes to people who had not witnessed its glitz and glamor directly. As the Cold War ended, the East looked toward the West and saw Hollywood. In his 2004 book *Why Globalization Works*, Martin Wolf, editor in chief of the London *Financial Times,* uses the literary reference "Oliver Twist Has a Television Set" to explain the power of image for creating a desire in population segments to have what they see in the lifestyle reflected and promoted by media. It was this timing of cultural international information exchange that provided the right environment for the proliferation of global brands.

For global brands to appear, they must speak to segments of consumers who share an ethos (Murray, 2005). Ethos denotes a sharing of foundational values, and for global branding it describes multicultural consumer segments that have a unique convergence of tastes. This convergence of tastes happens because of exposure to standardized information. As media become global, the standardization of their images amalgamates, and so do their promotional platforms.

As globalization integration increased, the promotion of brands to new customers escalated, and those customers responded with favorable readiness. Strizhakova et al. (2008) posit that global brands are well received in emerging and developing countries because they create an imagined global identity for people who had felt excluded or marginalized by living in a poorer nation. Anholt (2003) goes further, arguing that global brands can reinvigorate local cultures.

Building global brand equity relies on a unified promotional strategy. The face of the brand needs to be promoted based on global fame. As already discussed, Gianni Versace realized this fact when he focused on promoting the models and created the supermodels so that their faces became internationally branded. After his passing, creative director Donatella Versace—his famous sister—continued the legacy of global face awareness by refocusing the image promotion tactics of the company toward celebrities. Many fashion insiders questioned the future of the Versace brand after the tragic death of Gianni, and the company's sales suffered for 7 years until Jennifer Lopez stunned audiences at the 42nd Grammy Awards red carpet in a uniquely provocative dress by Donatella Versace. That dress became famous enough, as "the green Versace dress," to merit its own Wikipedia page, describing the design, occasion, and social and company impact of the image. It became the return of Versace, the reestablishment of the company's trendsetting power, and the validation to marketing professionals of the power of fame and celebrity in modern promotional platforms.

Celebrity product endorsement has been well documented and studied by the advertising, production management, and marketing fields for its value in strategic marketing communication (Erdogan, 1999). Celebrity impact has become increasingly important because of previously discussed technological platforms, whereby the image can be proliferated immediately, and its impact can be judged immediately as well.

With respect to fashion, as the story of "the green Versace dress" shows, neither designer, company, nor celebrity was prepared for the magnitude of impact of that particular image. The image's power made people react. From fashion enthusiasts, to music fans, to news anchors covering the event, everyone commented on the provocative cut, patterns, and fabric of the dress. Somewhat secondary in this process, fashion critics, stylists, and magazine editors had to chime in, but by the time their voice was offered a "microphone," it was already accepted by the fashion world that the dress was an iconic success. In contrast, in 1992, just 8 short years earlier, Gianni Versace had become infamous in a bold statement with his "bondage collection." Most fashion critics hated it, but the celebrities of the time embraced it as a sign of sexual liberation. That social message has made it iconic now, with homage paid to the entire collection by Lady Gaga, who has worn pieces of it in video, red carpet, and concert appearances since 2010. The difference between this modern-day example and that of past decades is that back then such transformative fashion statements as the 1992 Spring Versace collection (now known as the bondage collection) were made on the catwalk. The bondage collection images were promoted through the traditional print campaigns featuring supermodels Christy Turlington, Yasmeen Ghauri, and Naomi Campbell. The celebrity endorsement and embrace of the garments and look happened a year *after* their introduction and initial promotion. This was arguably the first time a sign was evident that celebrities could challenge the fashion critics. However, the challenge was a year-long battle of pugilistic promotion by Versace.

A greater sign of the power of shock through the immediate delivery of fashion statements away from the catwalk was again created by Versace, who, 2 years after the bondage collection, dressed unknown aspiring actress Elizabeth Hurley in what has been referred to as "the most iconic red carpet dress of all time" (see Khan, 2008)—the safety-pin dress—for the film premiere of her boyfriend (at the time) Hugh Grant's movie *Four Weddings and a Funeral*. Grant, the celebrity, was a mere sideshow to the impact his date made in the revealing and surprising dress, unbelievably held together by ostentatious safety pins. The dress poked fun of the whole concept of dressmaking, tailoring, and master stitching by removing the sewing from the process. Again, the critics hated it, but the media embraced it, replicated the image, and turned Hurley into an overnight global celebrity. The next step in cementing the celebrity as the new face of fashion, rather than the model, was once again pioneered by the house of Versace. Gianni hired superstar Madonna as the face of the Versace 1995 print campaign at a time when the employment of celebrities as models was not common. Today, celebrity models are not only common, but preferred.

In modern fashion promotion, a celebrity model spokesperson is essential for fashion-brand equity (Spry et al., 2011). As a matter of fact, one would be hard-pressed to find a fashion magazine cover nowadays that features a fashion model. Actresses are the only models placed on magazine covers today. First introduced by Versace, fashion guru Anna Wintour, often called the most powerful woman in fashion, embraced the idea of the celebrity model. Wintour took *American Vogue*, and by default the entire *Vogue* brand, down the path of celebrity models.

Versace and Wintour understood that in the new IT age, celebrities added more to a brand image than just a beautiful exterior. They offered an option of relatability. Whereas in the past models evoked feelings of aspiration, today celebrities add a social cause element to promotion that enables promoting a much deeper and more emotional connection to the fashion message. This message incorporates the whole legacy of a celebrity.

For these reasons, fashion marketing experts agree that celebrities are the perfect spokespeople for fashion brands, because their personas are the ideal fit for integrated marketing strategies. To that effect, Chen et al. (2013) studied brand consistency in product and brand positioning—the strategy to place specific brands together in integrated marketing—and found that the synergy between select brands is carefully thought out in promotion tactics.

When engaging celebrities, advertisers consider what other brands the celebrity endorses. This celebrity *endorsement portfolio* (Kelting and Rice, 2013) is important for message consistency. This tactic puts an individual brand in a situational context. It is this context and its consistency that lead to building a positive brand image in the eye of the consumer. Therefore, a two-way promotional partnership develops, in which integrated marketing occurs continuously. It extends past the expiration of a particular advertising campaign. This is the case because of message distortion and absorption in advertising.

Message distortion is a problem when the consumer misunderstands the meaning of the message advertisers intend to convey. Message absorption is a problem when there are too many messages. The consumer is exposed to many of them at the same time and therefore, only remembers a few. The multitudes of messages transmitted in the modern-day promotion of consumer goods are quickly forgotten.

With multiple media information sources, an advertisement's impact on customer memory is diffused. Therefore, with respect to fashion promotion, the fame of the spokesperson is essential. When a person sees a celebrity endorsement, after the point of the initial visual impact, that person is most likely to remember the celebrity, but not what product was endorsed. Kelting and Rice (2013) describe that impact process as consumer memory of celebrity advertising. They find that consumer memory is negatively impacted by the breadth of products a celebrity is promoting, as well as the frequency of image exposure and change. When customers see the same celebrity in several advertisements in close proximity to each other, they tend to forget the individual brands. However, they remember an overall image as an impression of a look embodied in the celebrity.

As noted, the celebrity spokesperson today promotes a package of brands. The message is that those brands work together best, at least according to the famous spokesperson. This type of complementarity in promotion is not new in advertising; however, what is new is that the tactic has become global. The same campaigns featuring the same portfolio of brands, promoted by the same celebrity, are diffused across national borders. Therefore, Okonkwo (2007) notes, the celebrity model must have global appeal. This is an example of global branding not only of products but also of public figures.

Considered impractical and even impossible up until the 1990s, global branding, particularly multibranding and integrated marketing, is a fairly recent practice based on taste and preference convergence. Choi et al. (2005) examine the phenomenon in the context of celebrity endorsements in a cross-cultural format, and argue that mass media exposure mitigates the need to create differentiated product images, which was the traditional marketing mantra in international marketing. It was based on the assumption that different cultural factors in different nations dictate the need to create differentiated images for the promotion of goods. Choi et al. (2005) argue that today one global image is sufficient if its spokesperson has a high international social profile. The creation and maintenance of that profile rely on its social context, because high-impact brand spokespeople do not only promote products, but must show a track record of promoting social causes.

As social causes in human rights, environmental stewardship, peace activism, and alleviating global poverty become transnational, so do their champions. The social impact of celebrity models makes the current advertising environment very different from the legacies of its past. In the past, advertising was based on fantasy. Today, it is based on real-world impact. Such a drastic change is dictated by a change in economic behavior.

The field of political economy examines institutional factors, including globalization, that contribute to a change in the industrial psychology of consumers. A key finding in that field of study is that a critical mass of consent must be reached internationally in order to equalize the social importance of an issue (Jaquette, 1997; Marwell and Oliver, 1993). Reaching the critical mass of consent predicates a major cultural change. Mantzavinos (2001) argues that cultural change must precede market change. Based on that argument, it can be deduced that the major cultural change that altered the fashion market into a fast-fashion market was the convergence of global tastes and preferences.

GLOBAL TASTE AND PREFERENCE CONVERGENCE

Taste and preference convergence has been the focus of much attention and skepticism because it challenges the tradition of promotional management.* That tradition is based on the justification that local tastes differ greatly from area to area—be it region, country, or continent—and the differences are vested in cultural heritages. As neo-institutional economists, such as C. Manzavions (2001) and Douglas North (1990), argue, cultural changes happen slowly through marginal augmentation of legacies. Therefore, as the field of advertising developed, particularly in fashion, campaign design was based on the foundational understanding that successful promotion happened when a marketer correctly identified the specifically unique taste and preferences in the target market. Then, successful advertising effectively tailored the promotional nature of its message that the product satisfied a core local taste need. It was assumed that those core taste needs varied because of cultural differences. However, as times have changed, scholars and advertising professionals are beginning to examine factors of cross-cultural convergence that have resulted from the increased communication across borders in the age of globalization. With that point in mind, Ozsomer and Altaras (2008) bring together the

---

* For a thorough review of academic research on the subject, refer to Whitelock and Fastoso (2007).

findings from the development of consumer theory and signaling theory (which studies advertising messaging) and put their assumptions to the test through examining networking association in consumer behavior. The authors explain that core cultural changes have occurred, whereby consumers look at brands for building their own cultural capital. This argument is contrary to the legacy of assumptions on branding and culture that argue the opposite causal relationship. It is assumed that culture builds brand capital, not the other way around. However, Ozsomer and Altaras (2008) are among the scholars who argue that a change in direction has occurred, and that change is evident across countries. In that context, global brands are found to be important because of their perceived higher level of credibility and realness.

To reach a level of consumer confidence that enables putting a brand in charge of building cultural capital that leads to thinning local cultural characteristics and replacing them with global preference convergence, corporations rely on what has been described as international brand architecture (Douglas et al., 2001). This is the coordinated proliferation of multiple products under the same brand that are simultaneously sold in multiple markets.

Multiple products under fewer brands are most effectively diffused if tastes are similar across market segments. It is this fact that is much studied in international marketing today. It is all based on the question of whether tastes of different peoples in different countries are indeed converging, and whether that convergence has reached a critical mass of similarity. Of interest, of course, is what factors build convergence.

O'Guinn et al. (2014) explain that integrated brand promotion builds convergence. The authors identify social media as a key component in integrated branding. Among the main reasons for this fact are the immediacy of message delivery, and also the presence of customer input. Social media promotion allows a global unification of attribute description, because the input on product attributes is from a diverse customer base. No matter how diverse, there is only a finite character space for description, and attribute description is finalized in an entity of common agreement of consumer input. It is this nature of building commonality that leads to taste convergence.

De Mooij (2013) explains that marketing and advertising theories' stress on segmentation is challenged by the popularity of ethnocentric approaches such as global branding. The explanation is put in the context of the question: "What constitutes good management in building an international brand architecture?" Traditional service and operations management assumptions would suggest that there is a higher need for segmentation because of the cultural strength of local customs. Therefore, with respect to building a solid international brand architecture, the conclusion is that the more internationalized a brand is, the higher level of segmentation it must possess. This conclusion has become foundational, as social science fields in general have taken a combative stance against Western ethnocentricity. Historically, advertising researchers have rallied against ethnocentric promotion, advising against unified brand strategies such as global branding. Among the reasons is the belief that customers appreciate the care of being understood. They respond well to feeling important, and having their local cultural legacies reflected in brand attributes indicates that the producers care intimately about their needs. Another reason is the assumption that because of spatial friction—the economic term for geographic and cultural distance among nations—there is significant esthetic separation in terms of tastes and preferences. However, it is evident that although cultural factors are highlighted across social science and behavior analyses as the core differentiation market factors, when it comes to consumption of branded goods, ethnocentric approaches work (Akram et al., 2011; De Mooij, 2013; Schuiling and Kapferer, 2004). With respect to fashion, ethnocentric approaches in promotion are particularly successful today because fashion advertising teaches buyers a new consumer culture. Tastes and preferences converge, because in fashion the brand promotion tactics dictate taste development.

Not just in fashion, but in general consumer behavior research, fairly early in what has been described as the global brand proliferation era (the years from the late 1980s to today), Ger and Belk (1996) explain that there is a global consumer culture in development. Collins (2002) concurs, and adds that it is based on a transnational economic and political interconnectedness. The building of this new global culture is based on the dominance of a few transnational corporations that manufacture and market the majority of consumer goods around the world. Wang and Wang (2008) explain that this dominance is a product of lowering intrabrand competition, that is, the competition among brands owned by the same corporate parent, as well as interbrand competition, which is competition with brands external to the firm's corporate holdings. It all leads to collusion in global marketing, because the incentives are to limit the number of competitors individual firms face.

Today, lowering the number of market competitors at the firm level happens through brand proliferation. Brand proliferation is the spread of a brand globally. Brand proliferation is described in terms of strength. A brand's strength denotes how global it is and also how dominant it is in relation to competing brands. Brand strength is of the essence when studying brand proliferation because of the assumption that it can serve as a market-entry deterrent for competitors. The more dominant a brand is, the harder it is for new competitors to enter the market, because brand allegiance of the local consumers is strong, and it takes a lot of resources to convince customers to try, and ultimately switch to, new brands in their consumption patterns. However, this assumption only holds true if brand-specific differentiation is not too large and if customers perceive the products on the market as being close substitutes (Wang and Wang, 2008). These two incentives necessitate that customers do not demand too much differentiation. Therefore, to control what customers demand, at the firm level, the incentives are present to dictate needs, rather than respond to them.

As Ger and Belk (1996) show, it is the corporations that drive the convergence of taste and preferences through their marketing of products. Dholakia and Talukdar (2004) also show how global market integration accelerates what they call the *homogenization of consumer behavior* and provide evidence of consumption convergence in emerging markets. As already discussed, Strizhakova et al. (2008) explain that feelings of inadequacy, isolation, and cultural immateriality contribute to the embrace of global brands by consumers in the emerging markets of the developing world.

In the developed world, there is also evidence of taste and preference convergence (Aizenman and Brooks, 2008). Collins (2002) explains that it is due to media proliferation and the actual convergence of media platforms. Whereas in the past, media had been clearly separated into film, television, telecommunication, and press, all in their respective national provisional platforms, today all are melded into a wider *international connectedness* (Collins, 2002, p. 1). Echoing points that this international connectedness has a politically cohesive aspect, Van Ittersum and Wong (2010) examine how taste convergence influences preferences for the preservation of local cultural uniqueness with respect to brand allegiance. Their research indicates that when asked about the importance of preserving local brands as saviors of local economic platforms, consumers showed decided support for local production. However, when price was included as a trade-off, consumers favored the cheaper, global brands. These findings indicate that consumer tastes and preferences are very changeable and reactive to information and promotion external to geographic market segments. Above all, the findings indicate that price is valued more highly than cultural loyalty. These facts help explain why taste and preferences are converging, and why this convergence makes it easier for global brands to control consumer behavior.

With respect to fashion taste and preferences, convergence was clearly noted as early as the 1990s (Bikhchandani et al., 1992). The mere fact that trends go global indicates that a critical level of convergence exists. With the advent of fast fashion and the proliferation of global branding, fashion taste convergence is accepted as a given. Therefore, advertising relies on this fact, and as Bruce and Daly (2006) argue, the nature of fashion advertising has been changed to reflect convergence.

In fashion marketing, research trend prediction has been impacted by this newly grown level of taste convergence. The influence of major fashion houses has diminished, and with it, the importance of trend-prediction agencies. In previous years, fashion retailers have relied on forecasting to assess the needs and wants of consumers. Trend-predicting agencies analyzed historical datasets and sold predictions on future preferable cuts, colors, fabrics, and accessories based on historical averages of successful sales. Today, they use real-time data. In the past, the trend-predicting analysis could start some 18 months before a product was to be sold (Hayes and Jones, 2006). Actual orders were placed with manufacturers a year prior to the time of sale. This process of product development through forecasting and starting the manufacturing cycle well in advance of the time of product unveiling is referred to as lead time in production. By the mid-2000s, Birnbaum (2008) argues that the lead-time process for fashion goods had shrunk from one calendar year to as little as 30 days. Modern-day apparel producers have only 30 days to create a look, manufacture it, and distribute it to their multiple internationally separated retail markets. Such lead times have demanded the proliferation of production and operations toward levels of integration that, much like modern-day products and tastes, defy nationality.

3

The Production Platforms of Modern Garment Manufacturing

INDUSTRY AGGLOMERATION

Lead times in production have shrunk because of the need for global fast response to the change in trends. Global fast response can only be achieved through relying on the production platforms of multinational corporations (MNCs). As a whole, MNCs dominate the apparel industry in all its facets (Gereffi, 1999). The reasons for this internationalization of the apparel production function have to do with trade policies* as well as the previously discussed trends in global brand proliferation. As a result of successful global brand proliferation, today all global branded retailers are MNCs. These retailers are able to extend their economic power by exerting control over prices through pressuring the independent labels they carry to compete on price by using their growing volume of private-label production as leverage (Miroux and Sauvant, 2005, p. 5). Branded retailers also achieve control over their suppliers by leveraging the services of new suppliers while maintaining a relationship with existing suppliers that have the capabilities to respond to fast change. This platform is referred to as category management (Sheridan et al., 2006).

Category management has been studied since the 1990s as a main innovation in retail operation because of the changing power balance from manufacturers to retailers. As Bruce and Daly (2006) explain, retailers have increased delivery pressures on manufacturers. Through category management, which at its core relies on information sharing across supply chains, retailers develop important relationships with, as Bae and May-Plumlee (2005) put it, "preferred suppliers." Seeking out specific suppliers

---

* To be discussed in detail in this chapter and in Chapter 4.

and maintaining an open information flow with them, even in aspects as far removed from the assembly process as trend predicting, is essential in modern fashion production because of frequent trend shifts. Fast-fashion retailers are especially reliant on the services of preferred suppliers, and for that reason, European fast-fashion MNCs such as Inditex invest in owning commercial interests throughout their entire supply chains.

The preferred suppliers of successful modern garment retailers need to be able to take larger repeat orders on a regular basis and have the capacity to accommodate frequent and often unforeseen change requirements based on trend shifts (Bae and May-Plumlee, 2005). Being able to accept large orders sporadically is not sufficient (Barns and Lea-Greenway, 2006). Therefore, the whole purchasing process has changed. Buying has become a strategic tactic, not just an operational necessity (Bruce et al., 2004).

Large retailers have large volume requirements, so they only consider large suppliers (Hale and Wills, 2005). This fact has led to the increasing role of foreign direct investment (FDI) as conglomerates are looking to expand capacity in emerging markets (Miroux and Sauvant, 2005). Producers in developing nations have limited financial and know-how capabilities; therefore, an increase has been reported in the foreign ownership of both textile mills and garment manufacturing facilities, particularly those that employ more than 1000 workers (Bruce and Daly, 2006). In relation to size of manufacturing facilities, Pan and Holland (2006) explain that garment manufacturing facilities are considered "large" if they employ between 700 and 3000 people, including offices and factories. Meanwhile, their textile suppliers are of "medium size," meaning that they employ between 40 and 200 employees between the offices and factories. The authors examine the dynamic in a clothing export cluster in what they describe as the "Greater China" region of Hong Kong, Taiwan and southeast China (Pan and Holland, 2006, p. 350). The cluster produces clothes for 50 global brands, and its economic activity centers itself largely on exports to Western markets. These distinctions in size are important because they reveal a very important fact about garment production today. Textile and garment manufacturing vary greatly in terms of labor/capital intensity. Garment manufacturing is labor intensive, that is, it employs many workers. Modern textile production, however, is capital intensive, that is, it employs few workers but requires expensive equipment. The dynamic of textile production is fully explored in Chapters 4 and 5.

In both textile and garment production, growth of capacity is dependent on the ability to attract capital. Fast turnaround is required in an integrated manner for a producer to meet such demands. A drastic reduction is necessary in the length of time needed for the conversion of fiber to fabric and of fabric to garment, and the delivery of finished products to customers (Bae and May-Plumlee, 2005, p. 2). Such a response can only be achieved through efficient reorganization of entire supply chains (Miroux and Sauvant, 2005). The two main changes with respect to such reorganization from traditional to fast fashion are agglomeration and vertical integration.

Agglomeration is the geographic concentration of commercial activity. In production economics, agglomeration refers to the formation of industrial clusters (Lazzeretti et al., 2014; Markusen, 1996). However in fashion retailing, agglomeration can mean any aggrandization of operations with respect to size. The size of firms impacts on the size of the orders they place. Size of orders has become the main capacity feature of fast-fashion business, because agglomeration of retail outlets—the proliferation of many stores in many geographic locations under the same brand ownership—dictates the necessity to place large orders of products that can be put into all these stores at speed and on repeated schedules. For these reasons, competition among producers has shifted from a process centered on price to one based on fast response (Mikic et al., 2008).

In garment manufacturing, agglomeration also has geographic features. When a firm is internationalizing its assets, choices are made as to the location of value-adding and cost-saving assets. Among the many reasons for this strategic dispersion are proximity to customers and the establishment of foundations for future growth. Because of this focus on future growth, emerging markets' potential is essential when making capital decisions. Decisions about building expensive textile plants, for example, take into consideration the strategic return on investment of a facility: that is, its capability to remain operational and profitable for years to come. Geographically, fashion sales are remarkably concentrated. There are three main markets—Europe, North America, and Asia—where two-thirds of all consumption is in Europe and North America, a quarter of total sales in Asia,* and merely a tenth of global sales in the rest of the world (Birnbaum, 2008).

That "rest of the world," as Birnbaum (2008) puts it, constitutes a pretty sizable geographic area—all of South America, Australia, and the entire

---

* Mostly in Japan, but also post-2008—the year of the data summations in the above discussion—much of the global growth in apparel sales is in the rapidly growing Chinese market.

African continent.* However, to date, apparel sales in these areas are minimal. This fact is a testament to the inequality of economic development globally, because, almost exclusively, all South American and African nations are underdeveloped. This fact also has implications for strategic fashion product positioning. Sales in that part of the world might have been insignificant up until the mid-2000s, but future growth potential there is significant. The economic conditions that are building this potential are discussed in detail in Chapters 4, 6, and 7. At this point, it is important to note that many of the manufacturing activities of the apparel sector are being steered toward the developing world. Among the reasons for this is competition among garment manufacturers to establish commercial presence in emerging markets.

Commercial presence at the manufacturing level allows strategic future expansion at the retail level. Fashion market analysts who track future market-growth potential note that currently the fastest growth rates in apparel sales are in Asia, particularly in regions such as mainland China, Indonesia, and India, which in the recent past had not been important retail markets. Just a short decade ago, these regions were at the same level of underdevelopment as are African nations today (Crofton and Dopico, 2007; Moore and Birtwistle, 2004; Zhang and Kim, 2013). Today, the economic growth throughout the developing world provides incentives for apparel producers to think strategically with respect to future sales there.

Apart from the much-discussed economic progress of China, which drives the entire industrial development of Southeast Asia, an additional impetus for strategic expansion toward the East is the economic growth in the former Eastern bloc. As an example, Gereffi and Frederick (2010) explain that the fourth largest importer of apparel in the world is Russia, and it is only a percent or so, depending on fluctuation, behind Japan. A decade ago, Russia's economy did not allow such sales volumes. The fact that it has grown so much, so fast, is important as an example of strategic future expansion possibilities when put into the context of total apparel sales.

For fashion conglomerates, the leading sales markets are the European Union, accounting for 47% of all global apparel demand in 2008, then the United States at 24%, then Japan at 7%, and then Russia at 6% (Gereffi

---

* This discussion refers to the sale of new apparel items. The main reason that two continents are insignificant markets for new sales is that in both South America and Africa there is an extensive commerce of used apparel. The dynamics of that commercial sector are examined in Chapter 6 of this book.

and Frederick, 2010). The implication from these estimates on growth potential is that it is important for producers to be either in or close to those geographic markets because of the decisive need for speed in production and merchandise management.

Because future growth potential in apparel sales is diversifying away from its historic reliance on Western markets, it is important to examine the strategic asset internationalization of garment manufacturers. Traditional Western-branded apparel retailers such as Liz Claiborne and Nike were "born-global" (Cavusgil and Knight, 2009). "Born-global" is the concept of firm formation as the creation of a fluid international partnership of multiple investors, stakeholders, and supply chain intermediaries. Born-global firms do not, in general, own any of their direct production facilities (Knight and Cavusgil, 1996, 2004). In other words, they have no factories of their own, but place orders with independent manufacturers. The process is called subcontracting. This strategy enables maximum cost savings in two ways. One is not spending money on building facilities. The other is locating in geographical areas where direct production costs are the lowest. The born-global firms choose their subcontractors based on price; therefore, they can benefit from being able to place orders with facilities that offer the lowest prices. These facilities (as discussed in the rest of this chapter) are overwhelmingly in very poor nations.

From their outset, such modern new enterprises are not bound by national boundaries. This outcome is a product of globalization policies that have eased the costs of operating and acquiring stakes in international assets (Adams, 2008; Ward et al., 1999). The business platform allows swift market penetration through forming joint-venture partnerships and engaging in mergers and acquisitions (M&As). The strategy has been very cost effective, because it saves on expensive capital outlays such as building and maintaining factories. The strategy has been very successful for the traditional branded apparel retailers, and is now, arguably, the way to set up a commercial entity. Not only in retail, but in other sectors, the born-global model defines formation of new firms. Such large conglomerates as the telecommunications giants Logitec and Skype were created in this way.

However, unlike these companies, fast-fashion retailers grew following the traditional model of internationalization. That model follows the steps of the natural progression of a national corporation that has internationalized sufficiently and gone global (Johanson and Vahlne, 1977). Zara's internationalization seems to have followed this classic "stage model" by first establishing considerable size in its home nation of Spain, then entering geographically or culturally close markets before taking opportunities in more distant markets (Lopez and Fan, 2009). H&M also seems to have followed the traditional stage model. The company started in Sweden, grew to prominence in its parent-nation market, and then globalized operations. It is important to note that the firms revolutionizing the fashion retail sector were not products of the born-global model, because this indicates their legacy of vertical integration. That legacy has impacted their managerial culture, and evidence suggests that as they expand, they follow the model of their home-nation business environments, which support the corporate ownership of unified production chains.

In applied terms, the differences noted above mean that European economies not only allow but also provide incentives for fashion MNCs to own cotton farms, for example, as well as textile mills, garment production factories and retail outlets. In contrast, the American business model, based on specialization in production, that is, advocating outsourcing of operations and ownership to specialists in order to increase specialization and, therefore, efficiency, is based on a lower degree of vertical integration and a higher degree of outsourcing. Vertical integration for fashion conglomerates has been much discussed as an important competitive advantage the European fast-fashion houses have in their global proliferation strategies, because it allows unprecedented speed in product positioning. Today, it is generally accepted that the "fast" in fast fashion refers to two components of product development—fast production and fast response to market signals.

Operations management often praises Zara as the leader in fast response (Tokatli, 2008).* In relation to this, Lopez and Fan (2009) explain that Zara, and Inditex as its parent MNC, have three distinct operational advantages: (1) vertical integration to achieve a fast turnaround time; (2) use of franchise and joint ventures for rapid expansion; and (3) use of the store as the main tool for promotion, with low spending on advertising. Crofton and Dopico (2007) offer a thorough comparison between Inditex's business model and the branded retailer business models of the other conglomerates. One main difference is that most fashion firms handle design and sales but outsource manufacturing. In turn, the manufacturers rely on independent textile providers in sourcing the

---

* For operational discussion of speed in fast response competition among retailers, see Tokatli (2007), Tokatli et al. (2008), and Bruce and Daly (2006).

needed fabric, then independent manufacturers to complete the garment assembly, and another tier of independent distributor–brokers for the shipping and receiving of the goods in different nations. Inditex strives to keep all operations in house. This means that the company builds an arsenal of factories that it either owns or controls. Its key facilities are in garment production clusters in North Africa and Asia, as well as in relatively higher-cost production locales such as Germany and Italy. It still bases its final assembly for the main West European market on a network of more than 300 factories in Spain and Portugal (Crofton and Dopico, 2007; Lopez and Fan, 2009).

Vertical integration, therefore, is essential in fast fashion, because 30-day lead times demand a high level of synchronization between manufacturing and merchandising activities. These short lead times also make it unadvisable to have wide geographic dispersion in subcontractors. There is no time for apparel components to be shipped to several intermediaries in far-off places. For these reasons, in addition to the need to be able to promote the latest trends quickly, when fast-fashion merchandisers outsource production, they do so only in nations that are close to their home markets.*

The European market, which gave birth to fast fashion, has changed global sourcing patterns in favor of closer locales. Since the apparel sector in Europe is dominated by small and medium-sized enterprises (SMEs), West European manufacturers have to rely on outsourcing for capacity. However, global sourcing, which worked in the 1990s, is no longer an efficient strategy, because often retailers require garment replenishment mid-season for particularly popular items (Bruce and Daly, 2006). EU retailers require that merchandise be presented *floor-ready* on hangers and with stickers attached—an activity that can only be undertaken at short distances (Bruce and Daly, 2006). Manufacturers are responding by switching production from east Asia to North Africa, Eastern Europe, Turkey, and India (Bruce et al., 2004; Tokatli et al., 2008).

The European Union is the largest importer of textiles and garments, but much production sourcing and investment is done internally, mostly in Eastern Europe. For example, Gereffi and Frederick (2010) explain that for the EU-15 nations—Austria, Belgium, Denmark, Finland, France, Germany, Greece, Ireland, Italy, Luxembourg, Netherlands, Portugal,

---

* Or, as is the case in emerging markets, to the market designed for final apparel sale.

Spain, Sweden, and the United Kingdom*—most of their leading apparel suppliers (except for Hong Kong and China) receive either duty-free or preferential tariff treatment. Those suppliers are regional and linked to the EU-15 either through EU enlargement policies or through past colonial and postcolonial trade relationships, as is the case with North African Tunisia and Morocco, which are both part of the Euro-Mediterranean Partnership. The Eastern European nations of Romania, Bulgaria, Poland, Hungary, and Turkey are part of the EU-27 or EU Customs Union.

An example of the benefits such proximity brings in incentives for production expansion in apparel manufacturing comes from a 2005 United Nations study of FDI at the country level. In it, editors Miroux and Sauvant discuss how in 2004, 18 new FDI textile and apparel plants were constructed in Bulgaria. This level of construction puts the tiny Eastern European nation third in the number and value of apparel FDI globally, behind only China and the United States (Miroux and Sauvant, 2005). Historically, Bulgaria has not had a strong textile sector: the presence of the industry there is fairly recent. Its growth can be attributed to proximity to market benefits, but also to the importance of external funding.

The example of new textile plant construction in Bulgaria with external funding indicates a larger reality. The reality is that today most textile production is done by MNCs from developed nations that manufacture in developing nations. In the EU examples of Miroux and Sauvant (2005) and Gereffi and Frederick (2010), the fact that the preferential trade platforms provide incentives for location of facilities in nations with lower trade barriers for reaching the rich Western national markets is much discussed. It is also indicated that funding production in poorer nations is so important as to have defined trade liberalization policies in the industry since the 1995 Agreement on Textile and Clothing (ATC).

By 2005, developing countries produced half of all global textile exports and nearly three-quarters of global apparel exports (Miroux and Sauvant, 2005). In certain nations, such as India, Pakistan, and Bangladesh, the apparel industry, driven by foreign assets, has grown to such levels as to define their export intensity. Yang and Mlachila

---

* The EU-19 area countries are the EU-15 countries plus the Czech Republic, Hungary, Poland, and the Slovak Republic.

(2007) explain the trade intensity of apparel dependence in developing nations and offer the example of Bangladesh, where over 70% of all merchandise exports are in clothing and textiles. The other nations in which over 50% of exports are in apparel are Cambodia, Pakistan, Mauritius, Sri Lanka, and Tunisia. In these nations, employment intensity is disproportionately geared toward the poor and toward women; over 50% of all manufacturing jobs are in the apparel sector, and on average more than 90% of those employed in these jobs are women. The reasons are related to cost management and wage levels, but also to the nature of production in the gender-segregated cultures of such, mostly Muslim, developing nations. For these reasons, attention has been drawn to the issues of exploitation of women in relation to perpetrating poverty and inequality. These issues are still very strong, as the Bangladesh garment factory collapse tragedy of 2013 reminded the world. In April 2013, more than 1100 women died when the Rana Plaza building in Dakha, Bangladesh, collapsed. The building housed garment manufacturing contracting facilities for some of the world's largest retailers, such as American J.C. Penney's, Italian Benetton, British fast-fashion chain Primark, and Canadian conglomerate Loblaw, parent company of fast-fashion brand Joe Fresh.* The tragedy raised visibility around the issues of inequality, exploitation, and profitability in apparel manufacturing, because the outcome of the discussion centered on the cheap labor problem.

However, that is only one aspect of cost-minimizing tactics in global fashion retail. The less discussed issue is the high social cost of textile production and the fact that it is very different from garment assembly. It is geographically different, and, for certain countries, the industrial agglomeration of global apparel retailing has provided an opportunity to create a competitive advantage in textile production alone.

Agglomeration and vertical integration at the retail level have put strong capacity and cost pressures on textile manufacturers (Bruce and Daly, 2006; Bruce et al., 2004). These trends have increased the importance of MNCs not only in retail, but also in textile manufacturing in particular, as the sector is the most capital-intensive link in the whole apparel industry (Barns and Lea-Greenway, 2006; Birnbaum, 2005, 2008; Hutson et al., 2005).

---

* For a thorough discussion on the production interests of MNCs in factories such as the Rana complex, see O'Connor (2014).

CHANGING SUPPLY CHAINS

For certain countries, it is textile production and not apparel that defines exports. In Pakistan, one of the leading exporters of both textiles and apparel, textiles have grown to comprise over half of all merchandise exports. In India, apparel exports account for 55% of all export earnings. However, only about 12% of these exports are in the form of ready-made garments, so that 88% of exports classified under "apparel" are actually in the form of fabric (Chaturvedi and Nagpal, 2003). The other global leaders in textile exports are Nepal (16%), Macao (China) (12%), and Turkey (11%). India is on a par with Turkey, and despite its overall reliance on textile exports, it also only supplies 11% of global fabric sourced by international buyers (Miroux and Sauvant, 2005, p. 4).

Fifteen developing nations, including China, India, Pakistan, Bangladesh, Egypt, and Turkey, account for 90% of global textile exports and 80% of global clothing exports (Adhikari and Yamamoto, 2008). Among them, China has risen to be the leader in the industry and is referred to as "the tailor of the world" (Mikic et al., 2008; Pan et al., 2008). Today, Chinese textile MNCs are the largest in the world, but still over 34% of Chinese textile and apparel exports come from Chinese enterprises financed by foreign investors (Miroux and Sauvant, 2005). In Indonesia, 95% of textile mills are foreign owned (Robinson, 2008).

East Asian producers are adjusting to the increased competition from Eastern European and African nations. For example, Lu (2012) shows that Chinese clothing exports grew on average by 9.66%, gaining an additional 7.19% market share in the global apparel export from 2005 to 2009. Chinese and other east Asian producers are able to compete on capacity because of three main factors. First, they are becoming more adept at moving from simple commodity manufacturing to incorporating design and branding, offering a full-service delivery (Bruce and Daly, 2006; Bruce et al., 2004; Birnbaum, 2005, 2008; Hutson et al., 2005). Second, they benefit from their own growing markets and increasingly produce for local consumption, which creates an added incentive to grow in order to accommodate both export and domestic orders (Miroux and Sauvant, 2005). Third, they, in turn, aggressively invest in less-developed nations, mainly in Africa (Gibbon, 2003; Brautignam, 2008; Busse, 2010; Minot and Daniels, 2005; Miroux and Savant, 2005).

African nations, in particular, have become very important to textile producers. As well as their proximity to the main Western markets, the well-established, albeit at times turbulent, trade legacy with their former Western European colonizers, and their demographics of young citizens, African nations enjoy strong preferential trade ties with governments and multilateral organizations, with the World Bank and the International Monetary Fund (IMF) in the lead (Jauch and Traub-Merz, 2006). Fairly low on the overall developmental ladder, African nations have also been (for good or ill) left slightly behind in the global integration trend, but perhaps for that reason, they remained mostly unscathed by the global economic slowdown post-2008. If anything, their growth rates remained remarkably strong (Jacobs, 2011). Among the reasons for the growth is strong support from American and other Western governments through preferential trade agreements for specific nations. These preferential policies are driven by the general commitment of Western governments to provide poverty alleviation opportunities. In addition, because of the War on Terror and increasing fear of insurgence support in African countries, additional preferential trade platforms were put in place post-9/11. As a result of this stimulus, Jones and Williams (2012)* show that African nations, even those classified as "low income" and "fragile"—a classification for risk of harboring terrorists—had impressive annual GDP growth rates between 2009 and 2012, with projected growth rates for 2013 and beyond of increasing magnitude as the global economy continues to recover. Imports to American and other Western nations from those countries benefit from lower duty and tariff burdens than other developing nations. For these reasons, there are incentives for foreign investors to locate production there. With respect to the apparel industry, many of the firms pouring FDI into "low-income" and "fragile" African nations today are Asian MNCs, the majority of which are Chinese. Those investments are part of a carefully developed strategy by the Chinese government to increase its economic and geopolitical presence in Africa.

The importance of Africa and China can be seen simply through the volume of trade. While in 2003 China's investments in sub-Saharan Africa were around $3 billion, they had surpassed $100 billion by 2010 (Jacobs, 2011). As a result, a triangular relationship in the trade of textiles has developed, in which African cotton, mostly from Uganda and Tanzania, is shipped to China but then reimported as composite fibers into often

---

* Refer to pages 2, 3 and 4, and particularly table 1 on page 3.

Chinese-owned factories in Lesotho, South Africa, and Nigeria to be turned into ready-made garments, mostly for Western Europe, but also for the local wealthier African market (Brautignam, 2008; Busse, 2010; Miroux and Sauvant, 2005*). African cotton exports to China are growing at such levels that, Delpeuch (2007) argues, cotton exports to China are the defining revenue streams for GDP growth in Benin, Burkina Faso, and Mali.

Exports aside, Chinese and other Southeast Asian MNCs are directly developing the African apparel industry. For example, Lesotho's industry is owned primarily by Chinese and South African firms. Madagascar's growth in textile exports can be traced to Mauritian ownership, but the origins of Mauritius' industry come from French parentage and Chinese investment (Jauch and Traub-Merz, 2006). In 1982, after a specific round of trade restrictions under the scope of the then-active Multi Fiber Agreement (MFA) that introduced significant trade barriers to Hong Kong apparel exports to the United States, a major influx of FDI came into Mauritius from Hong Kong MNCs looking to locate in nations with preferential import arrangements with the United States (Joomun, 2006).† The same scenario played out in the Northern Mariana Islands, where South Korean textile MNCs developed the local spinning and weaving industry, also looking for quota locales during the late 1980s and 1990s that had easy access to the American market (Gereffi and Memedovic, 2003). The Northern Mariana Islands also offer an additional benefit as a legal United States territory. Any product manufactured there is classified as "made in the USA" under the product-sharing clause of the US import code. Following the recent trade liberalization rounds that started with the advent of the World Trade Organization in 1995 (which will be addressed in detail in Chapter 4), other African nations, such as Kenya, benefited from FDI coming from Asian MNCs. Qatari and Sri Lankan MNCs dominate Kenya's fairly new apparel sector, and most of those Qatari and Sri Lankan firms were born from South Korean, Japanese, and British investments.

African presence enables Asian MNCs to shorten supply chains and shrink response times for European and American clients, which is essential in fast fashion. Although only 6% of all FDI in apparel goes to Africa,

---

* Page 11 gives examples of ownership of assets in terms of major country holdings in the receiving African nations.
† Joomun (2006) offers a thorough historical track of the growth of the industry through foreign investment.

its growth rate, tied to east Asian MNCs in particular, has been impressive (Brautignam, 2008; Busse, 2010; Gibbon, 2003). Miroux and Sauvant (2005) explain that of all FDI outflows in textiles and apparel between 2002 and 2004, 35% had an Asian corporate parent. For the past several decades that parent typically would have been Japanese, but in 2002–2004, Taiwan, Turkey, Korea, Malaysia, and China surpassed Japan in total number of projects. Scholars often overlook the fact that many of today's largest textile MNCs come from the South, particularly Southeast Asia. Wrigley and Lowe (2007) explain that this growth of Southern wealth is referred to as "the supermarket revolution" of developing countries. In relation to this, Christopher and Towill (2001), in tracking the evolution of supply chain management, note that supply chains have morphed into a combination of lean and agile—meaning easily amendable in terms of volume and direction—in those markets where costs are the primary "winning criteria," as the authors put it. Then, they explain that there are different markets in developing nations, where the winning criterion is availability. Because these two paradigms define the business environment of developing nations, particularly those in Southeast Asia, investors are attracted to the region. This revolution, as Wrigley and Lowe (2007) put it, has been fueled by retail FDI and retail-driven procurement restructuring.

RETAILER-DRIVEN BUSINESS MODEL

Wrigley and Lowe (2007) also posit that the supermarket revolution in the developing world is related to the overall transformation of retailing. From its outset, traditional retailing had been a fragmented, local, unsophisticated business run by owner-operators—the essence of the "mom and pop" corner store concept. However, since the early 1970s, retailers have grown into large, global, technology-intensive, powerful, fast-growth corporations managing their own brands (Kumar, 1997). Because of the fact that retailers own brands, hence the classification "branded retailer," and because they are internationalized, De Chernatony et al. (1995) explain that they engage in "demand-supply" behavior. This behavior is dictatorial in nature. The retailers dictate their brands' consistent features and do not differentiate the products significantly in different international markets. The retailers promote the brand's cohesion and rely less on input from customers. In effect, the supply side of retailing, that is, producer input in dictating to customers what is to be sold and why, drives commercial behavior, as explained in Chapter 1.

In modern retailing, the demand side of supply chain management is becoming secondary to the supply-side factors that define retail profitability. Therefore, the supply-side factors define the modern industry business model as retail driven. Leslie (2002) explains that the retailer-driven nature of the industry is related to gender. The fact that most workers in the retail sector are women, while the whole industry markets to mostly women as customers, allows a unique commodity chain customization. Female retail workers share lifestyle commonalities with women situated at the different links in the production, advertising, and consumption interfaces in the fashion commodity chain. Because many of the economic issues women confront at the different interfaces of fashion production and consumption are similar, the sector is amenable to organization across the chain. This organization, Leslie (2002) explains, is at the core of the fact that the modern fashion industry is retailer driven.

Gender dominance is only one of the reasons why the fashion model is retailer driven. Another reason is the structure of modern-day retail itself. It is a multilocation system, that is, retailers have many outlets. Systems of such a decentralized, multilocation nature are prone to stock imbalances (Jiang and Anupindi, 2010). The operations management theory on stock imbalances states that there are two options for improving balance: customer-driven search (CDS) and retailer-driven search (RDS). CDS works when the customer demand is absorbed throughout the chain to indicate where demand is unmet. RDS works when merchandise management signals clogs based on unmoved merchandise, and that information is used to move product to lower price points.

In fashion, as explained in previous chapters, the nature of trend pricing that has led to the proliferation of fast fashion created a legacy of such a culture of merchandise management, which strongly dictates the flow of commerce. For that reason, as fashion brands proliferated globally, by definition, the industry became increasingly decentralized. The ever-increasing numbers of individual retail outlets under the same ownership rely on RDS for efficient inventory management in product placement of goods that have decreasing life cycles. In other words, when the shelf life of fashion and related goods is undefined because of the unpredictability of modern-day trendsetting, RDS defines the industrial flow of commerce, because there is no time for CDS information to be absorbed into efficient retail operations. As McGrath (2010) explains, when retailing happens under fast-moving conditions, retail business models change. And as Abecassis-Moedas (2006) notes, apparel is subject to change more than any other consumer good. Abecassis-Moedas further explains that the apparel retail commodity chain is buyer driven, which means that large retailers dictate the business flow to the apparel industry partners, which include brand-named merchandisers and trading companies in a variety of (mostly) exporting countries.

In buyer-driven commodity chains, design decisions are made at the retail level, even though the trend unpredictability of the business would suggest otherwise. Purvis et al. (2013) posit that this dynamic is due to the high frequency of purchasing. In other words, as explained in Chapter 1, as the fashion element of clothing products increases, their shelf life decreases (Forza and Vinelli, 2000). Design is forecasted forward, and the agility of production options is essential because of unforeseen factors that can introduce new trend elements. However, Doyle et al. (2006) explain that there is a need for balance in the agility quest, because stable and reliable supply chain partners define operational and financial stability. Therefore, the authors conclude, based on data from interviews with senior buyers from large UK-based fast-fashion retailers, that there is some discord between the tone of the literature and applied operational tactics. The literature focuses on the problem solving of unforeseen trend changes, therefore stressing the merits of agility. On the other hand, the senior buyers explain that stable supply chain choices define their daily decision-making because of the risk of the unknown.

When sourcing from new suppliers, buyers must face risks of quality, promptness in delivery, and financial strain. Therefore, when choosing new suppliers, the buyers interviewed by Doyle et al. (2006) noted that reputation within the industry was a key piece of information that they sought. The overall result of evaluation is that the quest for agility often falls behind the need for reliability. Bruce and Daly (2006) explain the reason for this dynamic by the fact that in fast fashion, capacity dictates choice of suppliers. Few suppliers are large enough to take on repeat large orders. Buyers for fast-fashion retailers, whose ordering needs grow as each retailer expands and opens new branches, look for sourcing options of suppliers that can accommodate the large numbers of individual pieces of clothing that orders comprise.

These strategies have been in place since the 1980s and have grown in reciprocity with the industry. In the 1980s and 1990s, American retailers championed quick response (QR) programs to promote responsiveness by reducing lead times, cutting inventories, and reducing risk (Tyler et al., 2006). Those programs increasingly relied on outsourcing and defined the nature of retail-driven supply chain management because they created tiers of production networks involving different countries. Gereffi (1999) explains the geographic structure of these *value chains* through the development characteristics of the countries involved in the different processes in production.

The value in those value chains is derived from many sources: from cost savings obtained via strategic order placement (or sourcing) for labor-intensive processes in countries where wages are low, to locating knowledge-intensive operations in countries where relevant labor skills and/ or capacity (e.g., specific knowledge of the labor force or infrastructure, such as telecommunication capabilities, energy and natural resource, and input components) offer different sets of values. This strategy of combining low labor costs with relevant skills and production capacity is defined as industrial upgrading. Tyler et al. (2006) explain that industrial upgrading starts by providing export-oriented assembly options (also known as original equipment manufacturer [OEM] production) in one nation, and then integrating that production "forward" in the production chain with greater service–activity links or original brand name manufacturing (OBM) in different nations. Gereffi (1999) defines the strategy as dictatorial in nature, because the branded manufacturers act as "strategic brokers" that link overseas factories with evolving product niches into the main consumer markets.

It is the unique combinations of high-value research, design, sales, marketing, and financial services that allow the retailers to amass special industrial knowledge. Individual suppliers seldom possess such an integrated knowledge of the entire industry. Today, those that are more vertically integrated may possess such expertise, as is much discussed in the strategic analysis of fast-fashion conglomerates such as Inditex. However, back in the era of the branded retailers, when the retail powerhouses were the mega "born-global" chains such as Liz Claiborne and Gap and the athletic megabrands such as Nike, Puma, and Adidas, those retailers' order schedules defined production demand. The high value derived from the fact that the marketing of their products dictated customer demand. Branded retailers' knowledge of market dynamics, with respect to forecasting, became the key information source for the entire structure of the apparel, accessories, and cosmetics and fragrance industries.

The end result contributes to the retailer-driven incentives in operations management, because retailers face an incentive to dictate with respect to forecasting. They can discount new trend information coming up from the CDS if supply chain options do not offer a feasible production platform, and offer an alternative solution that their preferred suppliers can accommodate. To that effect, retailers dictate to both the manufacturers and the end customers through their behavior as buyers. They act as both buyers and trendsetters, or trend approvers, to put it another way.

Another feature of the industry that contributes to its retailer-driven nature is product bundling. In the context of retailer-driven market behavior, product bundling occurs when retailers combine component goods produced by separate manufacturers (Bhargava, 2012). This is very much the case in large retail stores such as Macy's and Sears in the United States, Royal Ahold in the Netherlands, Carrefour in France, and Marks and Spencer in the United Kingdom. Product bundling builds brand alliances. Brand alliances are strategic links between several independent brands that retailers decide to promote in tandem. This *market-facing* tactic defines the retailer-driven nature of modern-day fashion promotion (Wigley and Provelengiou, 2011).

In fashion merchandising operations management, structures are described as "back-of-house" and "market-facing." Back-of-house operations include product design, supplier management and logistics, and sales channel management. Market-facing strategies include all aspects of promotion. Therefore, the operations management nature of the fashion industry is retailer driven, because independent retailers are the main decision makers that shape bundling information through experience, then transmit that information toward the "back-of-house" management structures. With respect to branding, market-facing strategies refer to retailing and promotion. It is at these market-facing interfaces that the essential knowledge in customer relations is built.

With all the abovementioned factors that contribute to the change of the overall apparel production model from customer driven to retailer driven, industry insiders argue that vertical integration is the key factor for the major change (Barns and Lea-Greenwood, 2006). Zara and its parent company Inditex define vertical integration—the concept of owning the majority of the links in the production chain. As explained in Chapter 1, Inditex SA, parent company of Zara, owns textile mills, garment manufacturing facilities, and retail outlets. It can produce the main value-adding inputs "in house," so to speak, and rely on outcontracting to a lesser degree. This strategy lowers its vulnerability to disruptions in supply chain management and decreases the inefficiencies that are inherent in outcontracting, which are mainly due to information, transportation, and quality control issues. For these reasons, vertical integration reduces costs for business in the supply chain. Vertical integration is achieved with ease in modern-day MNCs through M&As. The nature of contemporary MNC asset management lies in effectively augmenting the level of vertical integration among subsidiaries. The goal is to align cost-cutting benefits in all links of production and exert control over elasticity of the firm's profitability.

Hayes and Jones (2006) set out to examine factors that contribute to the profitability of fast- and non-fast-fashion retailers, and find no statistically significant difference between the two business models. However, they note that reduced inventory is more important to fast-fashion retailers. This fact denotes the importance of moving clothes fast off the rack. Non-fast-fashion retailers, such as Dolce and Gabbana, for example (who famously refuse to offer fast-fashion lines), build their profitability on other product lines in cosmetics, fragrance, and accessories. Fast-fashion retailers mainly focus on clothing.

In summary, competitive price pressures have intensified in the largest import markets with the growth of fast fashion (Mikic et al., 2008). The retailers in these markets are growing globally following the branded marketing model of established conglomerates such as Nike, Reebok, and Liz Claiborne that pioneered a unique commercial platform based on two main models: outcontracting and retailer-driven supply chain management. Taking the most efficient features of the branded retailer operational platforms, fast-fashion retailers add an additional component to their operational strategies. It is a higher degree of vertical integration at the production level.

Historically, at the manufacturing level, branded marketers relied on buyer-driven supply chains of outcontracting, in which they did not own any production facilities but outsourced all components of production (Gereffi, 1999). The branded retailers dictated how and where manufacturing occurred, as well as how much profit accrued in the individual production chain links. Although changed to a degree by the vertical integration tactics of fast-fashion retailers, the branded marketer model's foundations still dictate production platforms in apparel. These foundations fuel the buyer-driven nature of the business. Gereffi (2001) explains that marketers and merchandisers at the design and retail end of the production chain, as opposed to the manufacturers, hold the main leverage in buyer-driven industries.

Branded retailers grow globally through the proliferation of their flagship outlets, and they achieve that growth by relying on intermediaries. The geographic location of these intermediaries is of the essence. As the shelf life of apparel items shrinks, so do the lead times in their production and merchandising. For that reason, the fashion conglomerates of the world united in search of trade liberalization policies. The quest for increasing speed to market provided the impetus to change the policies that defined the regulatory environment of the international trade in clothes.

4

Regulatory Environment

DEREGULATING THE INTERNATIONAL TRADE IN CLOTHES

In every region of the world and in every industry, proximity to market has always played a crucial role (Gereffi, 1999). However, geography's important role in fashion differs from the reasons of most other industries. In apparel manufacturing, the quota system, or the Multi Fiber Agreement (MFA), dictated production location decisions. The MFA provided a preferential trade framework under which developed countries imposed import quotas on apparel goods from developing countries. This massive international accord was championed by the American textile sector and negotiated by the United States, with the support of Canada and the main country members of what was then the Euro Zone—the precursor to the European Union (EU).

The quotas were broken into specific categories of textiles and partially and fully assembled garments, and were classified depending on origin. When it was negotiated in the 1960s, the MFA was designed to offer protection to American textile producers from cheap Japanese imports. At the time, Japan's textile industry was growing significantly in the post-World War II economic reconstruction. That growth gave the name to a whole economic literature on the *flying geese model*, first named by Akamatsu (1961). Akamatsu noted how it was the technologies in textile spinning, weaving, and dyeing that provided technological seeds for the development of larger machine- and chemical-dependent industrial sectors before World War II. It was that same industry that grew first and most successfully in the postwar reconstruction to reach significant export potential.

Back in the 1950s, leading to the 1960s negotiations that directed the formal international accord rounds that established the MFA in 1974, the increasingly disparate duties and tariff measures America imposed against developing nations' textile producers worried international relations and stability specialists because of the precarious political realities in textile-exporting nations. During the Cold War, a power struggle existed for political as well as economic influence between America and the Soviet Union, and that struggle most directly impacted political regime building in former colonial countries in Africa, Asia, and, to a lesser degree, Latin America, which had just won their independence. On the path of restructuring their political systems, there were opportunities to reestablish trade partnerships, because people in the former colonies were looking for options to break free from the commercial legacies of their colonial past (Ganesan, 1999, 2001; Jewitt, 1995). New political activists in these nations promoted freedom and independence ideals through promising a break with their former "owners." In order to achieve that break, many nations looked toward augmenting their trade relationships with the United States, and the American government eagerly promoted partnerships through its support for structural adjustment programs implemented through the World Bank and the IMF—two multilateral organizations dedicated to reconstruction and development of transitioning countries that are mainly funded by the United States (Jauch and Traub-Merz, 2006).

In this political reality, the cooperation of American importers as the core customers for textile products of developing nations was of the essence. The direct beneficiaries were the retailers who profited from the favorable results of increased competition in apparel inputs, such as fabrics and trim, as prices decreased. Also, more variety became available in fabrics, such as silks and tropical wools, that could not be manufactured in the United States. American textile firms, on the other hand, were disproportionately hurt by the cheaper and more varied imports and aggressively lobbied for protection. A decades-long battle of sorts ensued between textile producers and garment manufacturers, in which both were fighting for different aspects of protection. Textile producers lobbied for trade-restricting policies, while garment manufacturers lobbied for trade liberalization policies. Textile producers were richer and better organized than retailers at the time, for it was before the era of the megabranded retailers as we know them today, and continued to win disproportionately large numbers of individual protection measures against specific textile imports from specific nations. The result was a chaos of sorts and an intense competition for the status of most favored nation, which developing countries kept vying for in order to qualify for preferential trade exemptions from quotas and tariffs. The process was also happening in Europe, and therefore, from the 1960s, even before the formal passing of the MFA legislation in 1974, Birnbaum (2005) states that all aspects of the international trade of apparel were dictated by what became termed *the rules of origin* criteria of the MFA.

At the time, the United States was the largest consumer market for textiles and apparels. Because of its size, it had significant influence on world textile production, as it offered international apparel firms unprecedented profit opportunities. Total orders for US importers were larger than for any other single export market. However, at the time the United States also had a very large and established local apparel industry in all components of apparel production—from fiber spinning, to textile weaving, to fabric and garment manufacturing, to all leather, trim, and accessory components used in the final assembly of clothes. In the early 1960s, as the volumes of Japanese (and following the end of colonialism, other developing nations') textile exports grew, and these producers offered increasing variety, quality, and, most importantly, significant cost savings due to their cheaper production environments, American textile and garment manufacturers eventually united in lobbying for protection. As Akamatsu's flying geese economic model posits, the garment manufacturers, who had initially opposed trade restrictions, lured by the ability to source cheaper and more exotic foreign fabrics, saw how the textile industry was seeding garment production in textile-exporting nations, and eventually their suppliers became competitors, vying for manufacturing orders that normally would have gone to an American producer. Unified in their call to restrict free trade in the sector, American textile and garment producers helped create a system in which the United States imposed import quotas on textiles and apparel categories from 46 countries (Elbehri, 2004). European textile producers followed in joining the MFA, and the European legacy is that, until recently, the EU-15 (as defined in Chapter 3) had imposed import quotas on textiles and clothing from 21 countries (Elbehri, 2004).

From 1974, the MFA defined all aspects of the international trade of textiles and clothing. To that effect, the MFA, and its multitude of restricting preferential trade agreements that grew during the 1990s, created a network of production in specific locations (Gereffi, 1999; Hutson et al., 2005). However, as the retail sector changed, what became gradually more important for apparel conglomerates was access to international markets. The globalization and brand proliferation opportunities in international expansion, aided by the global market culture that developed in the 1980s and reached a maximum in the early 1990s, as discussed in Chapter 2, offered international market expansion opportunities that fashion strategists saw as very important for future growth. American market presence still defined profits, but future profit potential in the developing retail markets of the East provided a new set of incentives. These incentives changed from protection back to trade liberalization. As a result, on January 1, 1995, the MFA was replaced by the Agreement on Textiles and Clothing (ATC) to be implemented under the scope of the Textile Monitoring Body (TMB) of the World Trade Organization (WTO). In effect, the ATC was the structure for the gradual dismantling of the MFA. It was set up to implement distinct rounds of trade liberalization that would gradually ease Western market protectionist measures while allowing time for Western producers to adjust and cope with the new market pressures. The ATC created a slightly different network of trade partnerships from the one that had developed under the MFA, in which many new business establishments appeared in Latin American nations at the same time as the North American Free Trade Agreement (NAFTA) made Mexico the first developed nation with quota-free access to the American market. This fact provided incentives for creating new enterprises in nations that had established trade relationships with Mexico. The other main beneficiaries of the ATC were certain African nations (as explained in Chapter 3) and their funding parent conglomerates based in the Asian Dragon nations of Hong Kong, South Korea, Taiwan and Indonesia, and China.

Birnbaum (2008) discusses the fact that this network has remained fairly stable even since the end of MFA in 2005. Martin (2007) explains that its stability can be attributed to its very gradual removal and the economic interests of the major nations that trade in textiles and clothing. From 1995 to 2005, three distinct rounds of liberalization were supposed to lead to the complete removal of all protectionist measures in the international trade of all components of textiles, textile inputs, and finished garments. However, there is no such thing as complete removal of all protectionist measures in any trade for any sector. In the manufacturing of clothing, protection and export support policies, in the form of subsidies, are remarkably resilient. Martin (2007) explains how in the United States, with each of the distinct rounds of gradual tariff removal for most categories of clothing and textile components, special exemptions from liberalization were modified and even increased in categories deemed too vulnerable to external competition. This was not only the case in the United States—special protections and outward processing trade (OPT) agreements also continued in Europe. The first to be liberalized were quotas for underutilized categories and for less traded components. It was not until the fourth round of gradual restriction removal that most quotas were lifted.* Despite this, other protectionist measures, such as tariffs and rule-of-origin requirements, remain, affecting 49% of traded garment and textile components (Martin, 2007). Even after the removal of the quota system, China and the United States negotiated a memorandum of understanding that extended US market quotas for most Chinese clothing imports until the end of 2008. The EU also implemented separate temporary quotas (Wang et al., 2011). After the expiration of the temporary safeguard measures, Western apparel importers have resorted to antidumping measures (Martin, 2007; Wang et al., 2011). Antidumping legislation allows individual countries to impose trade-restricting countermeasures, such as duties, on goods that are believed either to be sold in the export market at prices that are below the home-nation market price for these goods, or to have been produced at below-market costs. By definition, these criteria are impossible to establish in apparel, particularly apparel components, because (as explained in Chapter 3) the production process is so fragmented that there is never a clear "home market." Furthermore, because most of the manufacturing is done in countries, such as China, where government-owned MNCs execute production, government price support in all cases mitigates variable production costs. Therefore, it can always be argued that government subsidies lead to lower-priced exports from these nations. Because of these facts, there has been an increase in US companies asking the US government to impose countervailing duties on imported textile products after the official end of the 2008 temporary and voluntary new quota system.[†]

Outright quotas are only one part of the story in protection of garment and textile producers from cheaper foreign imports. Preferential trade agreements, also referred to as trade preference programs, are arguably the decisive barrier to free trade and the defining factor in production location for new textile facilities to date.

---

* Martin (2007) defines four rounds of liberalization, because he includes the round that transformed the MFA into the ATC. The ATC itself had three rounds of separate policy implementations.
† This fact is stated by Reichard (2009).

During the quota years, production location for textile manufacturing facilities was dictated by the system (Miroux and Sauvant, 2005, p. 14). As the retail market globalized, so did the structures of manufacturing facilities. Foreign investment and multinational corporations defined the sector. Gereffi (1999) shows how MNCs began to dominate the whole apparel sector. Among the reasons for this growing trend is the general industrialization of Southeast Asian nations, such as South Korea, Taiwan, and China, and the expansion of their firms during the quota years to locations with available quotas (Gereffi and Memedovic, 2003).

As globalization underwent a profound impetus in the 1990s, garment and textile manufacturers faced another incentive to internationalize, in addition to the general incentives of market development and cost management that applied to MNCs from all industrial sectors. That additional incentive was the search for locales where MFA quotas were available. Quotas became the deciding factor in new plant construction, as growing apparel MNCs had to locate in nations that had available quotas. Quotas became a precious commodity themselves, traded on their own exchange in Hong Kong, where their price was greatly influenced by the amount and number of trade restrictions won by Western textile producers. Quote price signals had little to do with market demand, but were most closely tied to diplomatic happenings. Lobbyists for the textile manufacturers were the source of quota information for the pricing of specific categories. As cheaper and more varied fabrics became available in nations that increased their specialization in textile manufacturing, Western garment manufacturers saw an increasing incentive to outsource fabrics (Birnbaum, 2005). Therefore, domestic textile producers increased their pleas for protection. The more successful they were, the smaller and more detailed the quota categories became. In this way, quotas even dictated garment design. For example, in the 1990s, cotton fishing vests became popular because quota category 359 was readily available. Wool cardigans fell out of style because quota 434/435 was too expensive (Birnbaum, 2008). These and more examples indicate the methodical and calculated nature of garment production. Such evidence provides further support for the claim that the industry is retail driven, rather than consumer driven. Production-specific conditions dictated product development in a retailer-driven model. In a consumer-driven model, customer preferences dictate product development. Because of these production restrictions, the quota system may even have contributed to the industrialization structure and magnitude of emerging markets. Producers from countries with restricted quotas jumped borders and, willingly or unwillingly, helped in the industrialization process of the nations where they were able to locate (Gereffi, 1999). At that time, countries that were very poor, did not trade much, had no industrial structure to develop any of the apparel industry production platforms, and were considered insignificant or bad places to do business, had available quotas. These nations were most helped by the escalating demand for clothes in the West.

As the 1980s and 1990s amped up, and consumers in the West started to spend more and more on clothing, shoes, accessories, and cosmetics, as a result of the advent and promotion of global brands, the need to import more and more apparel increased. Therefore, nations that already exported heavily saw their quotas deplete at rapid rates. Then, their firms went global in the search for available quotas. Korean firms set up shop in Bangladesh, the Caribbean, and sub-Saharan Africa. Chinese companies expanded in Southeast Asia and Africa. Indian firms went to Nepal (Adhikari and Yamamoto, 2007). These decisions were driven by the need to find countries that had quota-free access to the American market, which, as noted, was the largest market at the time. Individual European nations, even with what was the European Common Market, had their own restrictions in place. Therefore, for exporters, securing a European retail client guaranteed a much smaller volume of sales than one of the national American retailers.

Access to the US market meant a volume of sales that at the time could not be matched by other market competitors.

Both in terms of sheer number of retail outlets and in terms of price, American textile and apparel importers could offer contracts that dwarfed any other national player. Therefore, the fashion power brokers were a remarkably concentrated group. The BBC's documentary series *The Look* estimated that in the 1990s, only about 400 buyers were invited to the fashion week shows. Most of them were from North America. In comparison, an average of 1600 fashion editors and journalists attended the shows. The selling of fashion print used to take almost three times the amount of input as the actual sale of the garments advertised in it. The largest buyers, those that could place the largest orders, dominated that sale. They were courted relentlessly because of their ability to place large orders *repeatedly*. Each of the eight seasons dictated that those buyers procure collections eight times a year. Producers that established a good rapport with a buyer had a higher chance of securing not just one, but eight, orders for the year. The old adage of "how much business you can throw my way" defined the level of a particular importer's power. For these reasons, the branded retailers such as Liz Claiborne, Sears, and Macy's, and the regional retail chains in North American malls, such as Belk in the southern United States or The Bay in Canada, were the desired customers. The more retail outlets the chain had, the larger the order its buyer could place, and the more important that buyer became.

At that time, retail chains were still mostly national. It was the luxury brands that went global by opening their boutiques or flagship stores in different countries. Still, their proliferation was mostly in mature markets, and, once again, their goal was for the highest expansion in the United States. The United States was not just the largest market but also the richest market in terms of disposable and discretionary spending. The average American consumed far more apparel than any citizen in any of the other industrialized nations.

Based on these dynamics, by the mid-2000s, Miroux and Sauvant (2005) estimated that in terms of market size the EU was the largest common market in the world, but much of the production there is sourced within the borders of mostly newly admitted members. Puig et al. (2013) explain that textile and clothing manufacturing activity is concentrated not only in new member states, but also in Spain, Portugal, and Italy. Their study offers a risk assessment analysis for survival in a turbulent and competitive environment. Because of government support, these nations had been able to maintain textile and clothing clusters, with textile manufacturing being mostly centered on yarns, knitwear, and specialized textiles—by definition, more expensive than "yard fabric"—the industrial description of finished woven fabric that has broad (mass-market) use.

Even the specialization in higher-priced products has not been able to shield these clusters from international competition. Puig et al. (2013) examine dismissals, bankruptcy, and insolvency proceedings that have led to factories closing, and find that one way to mitigate the international competitiveness risk is to focus on regional spatial expansions in locales that allow the fastest order response. European manufacturers remain competitive because of proximity to their main retail clients and can offer such deliverables as overnight trucking of finished garments completed with labels and stickers or wire racks (Barns and Lea-Greenway, 2006). Such finished-product deliveries are aided by sourcing from textile factories close by.

For these reasons, the EU market is the largest in volume, but when it comes to imports, Miroux and Sauvant (2005) estimate that the American market remains larger by 11%. Because of its market size, preferential trade with the United States was essential for nations reliant on clothing exports. When Mexico was given quota-free access to the American market via NAFTA, its textile and garment market share in the United States increased from 2% to 10%—a 500% jump in less than 5 years (Birnbaum, 2005). In the 1990s, six countries from what would become the Central American Free Trade Agreement (CAFTA) increased their US market share from 6% to 15%. Even countries without any previously existing garment industries, such as Jordan and the beneficiaries of the African Growth and Opportunity Act (AGOA), showed rapid growth from the moment they were granted quota-free access to the American market (Birnbaum, 2005).

In America, textile firms and domestic garment manufacturers both lobby for protection from imports. They have battled each other for years. Textile firms are better organized and had won protection from imported fabrics. This fact accelerated the decline of US textile manufacturing, because domestic factories were not allowed to benefit from the variety of fabrics available from foreign suppliers. That particular protection remained until 2005, but according to Abdelnour and Peterson (2007), most of the restrictions on the importation of foreign fabrics have been lifted. For these reasons, during the quota years, countries such as Jordan saw their garment industry grow—they had quota-free access to the US market, and fabric imports were allowed.

The US government used clothing quotas very strategically after the tragic events of September 11, 2001. In order to expand terrorist-hunting operations in Pakistan, the US government succumbed to pressures to negotiate quota amounts for Pakistani garments and textiles. Pakistan was then, and remains today, among the largest textile and clothing exporters. At that time, its economy was particularly dependent on the clothing sectors. It accounted for 1.2% of global apparel exports; however, over 70% of all Pakistani exports were comprised of apparel, of which textiles accounted for close to 48%, while finished garments accounted for 23% (Miroux and Sauvant, 2005). These figures explain why textiles are such a sensitive sector. It is worth examining why certain nations, such as Pakistan, become net textile exporters, while others, such as Lesotho, become net finished-garment exporters. Since the quota years, some things have changed in terms of market power and major import-market global balance.

According to the WTO's 2013 International Trade Statistics database, in 2012, the United States and the EU held equal shares of the global textile import market. The EU accounted for 20.3% of global imports from outside the Union. However, if intraunion trade is counted, over 38% of all clothing imports are directed toward continental Europe. The United States accounted for 19.9%. The data also show that between 2005 and 2012, the EU increased its clothing imports from within and outside the EU at a rate averaging 4.5% a year, while the United States only increased its import volumes by 1%.

These data explain that the dynamics of apparel purchasing at the retail and wholesale levels have been impacted. The European market's power can be attributed to the escalation of fast-fashion retailing. Fast fashion was born in Europe, and fast-fashion operations management is being perfected there, leading to increasing sales volumes at the retail level due to the promotion, proliferation, and diversification of fast-fashion product lines in continental Europe. In addition, European accession—the adding of new member states to the EU—has offered the EU its own internal emerging market. In the new member nations, former members of the Eastern Bloc, economic growth rates have been averaging 6%–8%, even during the global recession (Aslund, 2012).

Retail sales of apparel grow substantially with economic growth, and in the former Eastern Bloc, these sales are particularly strong because of the status symbol they represent. Buying Western brands is seen as a very strong indicator of social and economic success, and wearing the new trendy and hip brands is seen as an actualization of proof of global citizenship, denoting worldly sophistication and higher cultural status (Strizhakova et al., 2008). However, this actuality is also contingent on price. Nowhere is low retail price more important than in less developed, although more rapidly developing than their developed neighbor, nations. When median incomes are low, prices need to be within the discretionary spending brackets of those incomes. If retail prices need to be low, production costs need to be much lower. Therefore, most production growth has been in the poorest of nations.

POLICY WINNERS AND LOSERS

Production happens in several stages, during which transportation and distribution of components and finished garments happen in specific specialized geographically linked areas. Often, components—from bales of raw cotton, to woven fibers, to finished fabrics, to partially assembled clothes, to fully assembled clothes with trim (button, zippers, laces, and labels)—are repeatedly shipped back and forth across the same borders. This phenomenon is referred to as triangle manufacturing (Lu, 2012). The dynamic is global, but is best illustrated by what Jacobs (2011, p. 37) terms "triangular trade in textiles" between China, Africa, and the West. Cotton exports from Uganda and Tanzania, for example, are processed into fibers in Chinese textile mills, then the finished yarns and fabrics are exported back to Africa, but usually to other countries than the cotton exporters from where the bales were originally imported. The major African garment-producing nations are South Africa, Lesotho, and Nigeria. The garments sewn there are for the Western market.

Most of the companies engaged in the final assembly in Africa are, indeed, Chinese firms that had opened affiliates in Africa to take advantage of preferential trade agreements (PTAs) such as AGOA (Jacobs, 2011). The Act was passed by the US Congress in 2000 in order to stimulate more Afro-American trade by easing trade restrictions. In the case of garments and textiles, it allowed certain nations quota-free access to the US market. To qualify for such benefits, the rules-of-origin provisions of the act delineate that the garment is labeled as produced in a particular nation. The rules say nothing about the nationality of the firm that is manufacturing the garment. This is the case with all PTAs—they offer easing of trade rules based on nation of origin of final assembly, not on nation of origin of the parent corporate holdings.

Some PTAs are more restrictive than others, such as the European Everything But Arms (EBA) agreement (Asche and Schüller, 2008). The EBA is part of the EU's General System of Preferences (GSP), which has codes that are significantly more complex in textile and clothing regulatory compliance, but they mostly deal with the safety of the final product and regulate the use of certain textile dyes that have been found to be harmful. When it comes to final integrated assembly, apparel supply chains are shaped based on rules-of-national-origin import codes. Because of this platform, fairly long and complicated supply chains have developed in this triangular trade. Figure 4.1 illustrates the first-stage processes and their links in the overall industrial supply chain. These are the stages of fabric and garment manufacturing.

Often, fabric sourcing happens in several different nations (Gereffi and Memedovic, 2003). The process involves the sourcing of inputs for the

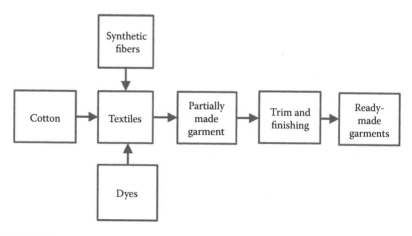

FIGURE 4.1
Flowchart showing supply chain links in fabric and garment production.

spinning of yarns and their processing, dyeing, and weaving into fabric. In certain nations, such as China, India, and Pakistan, sourcing of the entire process can be internal, as these countries have large, vertically integrated apparel industries. There are a variety of textile mills as well as garment factories within their borders. The same can be said about Spain and Portugal, with respect to production for niche European markets. However, usually fabric is sourced and shipped to a different country for assembly. This is the case because during the quota years, partial country-of-origin inputs dictated volume of orders. In other words, large orders that could be sourced entirely in India, for example, could not be sold in the United States because they would exhaust the quota limit for country-of-origin requirements in specific garment categories. For these reasons, incentives existed (and still exist) to source the fabric from India and then ship it for assembly to another nation where quotas for the particular garment category were available. Figure 4.2 illustrates the second-stage

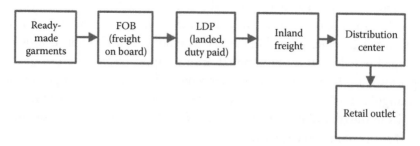

FIGURE 4.2
Flowchart showing supply chain links in the commerce of finished garments.

processes and their links in the overall industrial supply chain. These are the stages of finished-garment merchandising and distribution.

Because of these incentives, international specialization in final assembly occurred, whereby individual countries became specialization hubs in individual production links. The leading fabric exporters became India, Nepal, Macao, and Turkey (Miroux and Sauvant, 2005). The leading garment assembly centers became Bangladesh, Sri Lanka, and Vietnam (Audet, 2004), and more recently Myanmar, Cambodia, and Tunisia. Assembly there tended to be described as "partial assembly" because the finishing processes, which included adding the trim (buttons, zippers, toggles, and the like), labeling, and packaging, happened in the nations that would bear the "made in" label. These finishing centers are in Hong Kong, China, South Korea (Gereffi and Frederick, 2010), and also, to a lesser degree, Turkey, Egypt, Madagascar, and certain Latin American countries, such as Haiti and Honduras. Gereffi (2006, p. 35) explains that during the MFA, because of the limited nature of specific quotas, North American and European textile and apparel markets received imports from "50 to 60 different developing economies."

The geographic dispersion of the supply chain is a legacy of both MFA and the remaining preferential trade agreements. Preferential agreement systems exist in many places. In the United States, the system known as 807/9802 allows production sharing with facilities located mainly in Mexico, Central America, and the Caribbean. In Europe, the phenomenon is referred to as OPT, with partners in North Africa and Eastern Europe. In Asia, Hong Kong is the center, with outward processing agreements tied most closely to mainland China (Gereffi, 1999). When the quota system ended, PTAs remained.

The rationale for PTAs is that they make it possible to minimize production costs locally while developing a widening circle of exporters and intermediaries external to interregional supply chains. However, PTAs can have negative consequences for competitiveness. When NAFTA and the EU created preferential trade blocs to promote a growing consolidation of supply chains within each region, China and India learned how to be more competitive by aggressively increasing consolidation and vertical integration (Gereffi, 1999). PTAs have hurt some countries because the removal of the MFA ended the incentives to source from those countries. PTAs required that a certain percentage of the final product be made from the preferential partner's product, usually yarns or dyes from the United States and the EU, which are relatively expensive. Because PTAs demand that producers meet rules-of-origin requirements, costs have become too high for lower-developed-country (LDC) producers to remain competitive. Also, PTAs require preference-receiving countries not to use textiles or other inputs from competitors such as China. The result has been that such measures have offered protection to very few Western manufacturers.

This captive-market strategy has also posed a challenge of capacity. Since most Western exporters of textiles are small in capacity and inflexible in the variety of textiles offered, requiring LDCs to tie procurement to them also limits their flexibility in product breadth. It also limits LDC producers' capacity capabilities, regardless of willingness to expand. This problem is the biggest challenge for LDC nations that are part of US and EU PTAs, as they cannot compete with the capacity and flexibility of Chinese and Southeast Asian producers. The end of the quota system has impacted nations such as Morocco, Rumania, and Tunisia, where export volumes have dropped and the industry has started to shrink (Adhikari and Yamamoto, 2008).

A few countries that have both high productivity and low costs have benefited, particularly China, India, and Vietnam. Nations that are challenged as a result of its removal include Organisation for Economic Co-operation and Development (OECD) members and small country producers, largely because of capacity constraints (Mikic et al., 2008). Capacity refers to the ability of local factories to undertake ever-increasing order volumes. From the beginning of the quota phase until its end, world clothing exports alone increased from $158 billion to $276 billion (Lu, 2012).

When the system ended, so, conspicuously, did the research on its impact. Not until 2012 does Lu argue that in terms of spatial management, it is still unclear how the international trade of apparel components has changed. However, a clear pattern of change in market share is noted. Post-MFA, it was the European countries that achieved the fastest export growth from 2005 to 2009, and overall had the largest global market-share gain from 2000 to 2009 (Lu, 2012). This fact is very important because it is a result of empirical analysis of 51 nations, ranked in order of clothing exports in relation to all their merchandise exports. Undoubtedly, the growth rate observed is inflated in their favor because of inter-EU exports and trade. Still, it is important to note the overall volume of trade, because inter-EU production is contingent on imports of components from outside the EU, as none of the EU nations are leading textile or fiber producers. The one nation that is, is China. Lu's (2012) data shows that from 2000 to 2009, China achieved more global market-share gain than any other single nation. It was also the only nation that noted a significant increase (7.9%) in total market share of exports. Birnbaum (2008) explains that China has become the "tailor of the world" since the phasing out of quotas in 2005, because it has internalized its garment production.

China grows its own cotton, processes it, and is poised to take on any manufacturing order large enough to accommodate not only Walmart, but also all the large retailers in the world. The reason is that while US manufacturers focused on specialization of production and outsourced all logistics to third parties, thinking that all they needed to worry about was manufacturing around their core competencies while the customer had to worry about all the rest, the Chinese manufacturers adopted the outlook that the customer's only worry should be to write the check. If a customer had to place even a relatively small order for a test product line with an American garment manufacturer, that customer would have to worry about getting the fabric to the factory, then freight on board (FOB). Most American factories accept only CMT (cut/make/trim) orders, where there is a surcharge for "T"—trim. In getting the fabric, the customer would have a very limited choice of type and color, and would find it impossible to locate a mill that would accept specialty dying and pattern orders. When the garments were ready, the customer would have to worry about getting the garments out of the factory, trucking them, and storing them. The Chinese factory, on the other hand, would do it all. It would find any fabric, any color, any pattern, and would accept any specs. It would provide a sample for approval, and then it would take on not only manufacturing, but FOB, landed, duty paid (LDP), and, in some cases, facilitation with local inland transportation companies in order to get the product to the right distribution center.

All other leading exporters saw their global market share barely change. However, significant increases in individual export growth rates are noted, particularly in smaller nations that are geographically close to the European Market. Table 4.1 shows the largest annual percentage increases in exports for those nations that seem to have benefited from the end of the MFA.*

As Miroux and Sauvant (2005) predicted, indeed, the largest "winners," apart from China, were the other Asian leaders in textile exports (India, Bangladesh, and Indonesia), their regional partners (Cambodia and Myanmar), and the smaller European and African nations with notable

---

* Data are compiled and classified based on the appendixes in Lu (2012).

TABLE 4.1

Countries That Saw Largest Annual Growth in Clothing Exports Post-MFA

Country	Annual Percentage Change in Apparel Export Growth, 2005–2009
Albania	10.21
Bosnia and Herzegovina	22.27
Egypt	13.4
Moldova	7.35
Bangladesh	9.32
Cambodia	7.45
India	7.44
Indonesia	4.51
Madagascar	5.78
Myanmar	11.13

proximity to the EU. However, the data do not seem to support the conclusions of Mikic et al. (2008), who argue that the end of MFA would most negatively impact OECD member nations and small country producers. Lu's (2012) study shows that small country producers benefited the most, as well as the EU member nations—all OECD members. The combined EU member-nation export growth rate was 3.2% between 2005 and 2009.

The biggest "losers" in terms of export growth were North and South American nations, that is, the United States and Canada and their main South American trading partners. Table 4.2 shows the percentage change in growth rate of the major nations that export apparel located in the

TABLE 4.2

Countries of the Americas Export Growth in Clothing

Country	Annual Percentage Change in Apparel Export Growth, 2005–2009
Canada	−14.27
Colombia	−10.03
Costa Rica	−19.96
Dominican Republic	−27.81
El Salvador	−7.56
Guatemala	−8.66
Haiti	2.34
Honduras	−3.93
Mexico	−13.11

Americas from 2005 to 2009. The data are based on the appendix in Lu (2012).

Only Haiti was able to increase its exports. Every other major exporter to the US market decreased volumes. Mexico's data are particularly telling because, as already discussed, under NAFTA it had become the largest exporter of finished garments to the United States. Birnbaum (2005) states that in 1998 alone, 4 years after the implementation of NAFTA, its growth had risen to such levels as to surpass both China's and Hong Kong's combined exports. However, with the finalizing of the quota phase-out time frame, garment producers started leaving Mexico. Between 2002 and 2003, 325 of Mexico's 1122 garment factories closed down and moved elsewhere, mostly to Asia, in anticipation of the end of the system. Many of these companies were owned by foreign investors (Ryder, 2003). The data from Myanmar also point to the importance of foreign ownership. In 2014, Gap made news as the first Western clothing conglomerate to enter Myanmar after decades of sanctions and years of postsanction institutional instability.* Among the analysis of news were notes that Gap was late in arrival, as Asian, mostly Thai, Chinese, and Malay, MNCs had established a strong presence in the growing Burmese apparel industry. They had been expanding in Myanmar despite Western sanctions because of its strategic location and, as Trindle (2014) explains, lack of labor and wage laws. As already discussed, the African textile and apparel sector is mostly foreign owned. It is not just small and relatively underdeveloped nations that rely on foreign investors for the growth of their garment industries. In Indonesia, one of the largest exporters of both textiles and finished garments, more than 95% of textile mills are foreign owned. Most of the Chinese textile mills are also products of joint-venture partnerships, many of them with American MNCs (Birnbaum, 2008).

RECUPERATING REGULATORY LOSSES THROUGH ENVIRONMENTAL SOURCING

The growth of foreign ownership has raised ethical and environmental concerns that textile MNCs are strategically locating in countries that are still developing environmental regulatory systems, such as Indonesia,

---

* For a discussion see Trindle (2014).

China, India, Bangladesh, and Vietnam, in order to exploit regulatory uncertainty (Greer et al., 2010; Khan et al., 2009). The incentives for environmental sourcing are intensified in heavily indebted countries because they are under pressure to pursue structural adjustment programs that promote export-oriented production (Easterly, 2005; McMichael, 2004; Winters, 2010). Shandra et al. (2008) examine the 50 poorest and most indebted countries between 1990 and 2000 and find strong evidence that their organic water pollution levels increased with the amount of structural adjustment loans. Schofer and Hironaka (2005) explain that this is the case in most LDCs because their weak institutional structures lead them to succumb to both internal and external pressures to lower environmental standards for export-oriented production in order to stay competitive and be able to make structural adjustment loan payments to the International Monetary Fund (IMF). Structural adjustment programs, Easterly (2005) explains, are implemented by the IMF and the World Bank. The governments of the individual countries must apply for the loans, and the impact of their success is judged mainly through the increase in national exports for these countries. As already discussed, in many of those countries, such as Pakistan, Lesotho, Indonesia, and Nepal, apparel exports constitute the majority of all merchandise exports.

International environmental impacts of industrialization are studied across disciplines. Most empirical evidence of industrial pollution from FDI has come from the field of sociology (Chase-Dunn, 1975; Grimes and Kentor, 2003; Jorgenson, 2003, 2007, 2009; Rice, 2007). The field has pioneered the concepts of ecostructural investment dependence, arguing that a large proportion of foreign investment in LDCs finances highly polluting and ecologically inefficient manufacturing processes and facilities that are outsourced from developed countries (Grimes and Kentor, 2003; Jorgenson, 2007; Lee, 2009a). Across academic disciplines, this charge is referred to as the pollution haven hypothesis, the race to the bottom phenomenon, or the theory of ecologically unequal exchange (Jorgenson, 2009; Gray, 2006; Ibrahim et al., 2008; Lee, 2009a; Pan et al., 2008; Rice, 2007; Roberts and Parks, 2007; Smarzynska and Wei, 2001; Tufekci et al., 2007). The theory has increasingly generated interest since the early 1990s, but earlier studies have found little empirical support (Elliott and Shimamoto, 2008). Supporting evidence comes from fairly recent cross-national studies of greenhouse-gas emissions and other forms of air and water pollution (Jorgenson, 2003, 2007, 2009; Kentor and Grimes, 2006; Shandra et al., 2008; Smarzynska and Wei, 2001; Wagner and Timmins,

2009). The main contention is that the globalization of commodity production has enabled developed countries to partially externalize their environmental costs to LDCs. There appears to be an incentive system under which developed countries favor terms of trade with their developing nation partners based on greater access to natural resources and sink capacity (Jorgenson, 2009). Sink capacity, or carbon sink capacity, refers to carbon sequestration or the process of transferring atmospheric CO_2 into the soil (Lal, 2004).

Scholars who study environmental issues contingent to the garment industry have paid special attention to cotton. Cotton is one of the most significant crops, second only to food grains in terms of value and volume among agricultural products (Pan et al., 2010). It is one of the most important fibers, accounting for more than half of all fibers used in clothing and household furniture, and totaling 38% of the world fiber market (Pan et al., 2008). The World Bank reports that 130 countries produce cotton, while 170 are involved in the cotton trade. Among them, only 16 nations had cotton export value in relation to GDP of over 1%, including such unstable and underdeveloped nations as Syria, Chad, Mali, Myanmar, and the Central African Republic (Fortucci, 2002). Among these, some were so dependent on cotton exports that the majority of all exports came from cotton. Fortucci (2002) estimates that in Benin, cotton exports account for over 90% of its national exports, in Togo over 70%, and in Mali over 60%. Still, these nations are tiny global exporters in relation to the largest exporters.

As estimated by Birnbaum (2008), the United States accounts for more than 45% of total raw cotton exports. At the same time, cotton exports only comprise less than half a percent of American exports (Baffes, 2011). Using data from Doom (2011), who offers raw numbers of exported metric tons, as reported by the US Department of Agriculture (USDA), the next four exporters in term of percentage of all exported cotton are India, at about 20%, then Australia, Uzbekistan, and Brazil, with about 5% each. Adding these averages, and it is important to note that they are gross estimates, it is evident that although 130 nations produce cotton, around 80% of global cotton exports come from only five nations. In each of those nations, except for Uzbekistan, cotton exports, as a percentage of all merchandise exports, comprise a negligible share. Meanwhile, the top consumers of cotton are China, India, Pakistan, Turkey, and Brazil. India and Brazil are also exporters, so their cotton production mostly satisfies domestic manufacturing needs, and the surplus is exported. However, China, Pakistan, and Turkey's domestic production falls significantly below their needs, so they become net importers of cotton. Then, they process it into fabric, some of which is used in local garment assembly, but some is exported in the form of textiles. In the end, the majority of finished apparel items are exported, mostly to the United States and the EU. This trade dynamic has raised issues of sustainability with respect to carbon footprint problems and regulatory compliance (Dem et al., 2007; Gibbon, 2003).

5

The Carbon Footprint of Textile
Manufacturing for Fast Fashion

THE ECOLOGICAL IMPACT OF FIBER
PRODUCTION AND SUPPLY

Weaving cotton into fabric is the strongest value-adding link in the production chain and the most expensive component of garment production (Birnbaum, 2008). When speaking of "value" in this context, it is understood that it is value derived from the direct, as they are known, inputs in the manufacturing process. As already discussed in previous chapters, one can argue that when it comes to clothes, the strongest value in terms of final price of a garment is added through the complex advertising processes of branding, marketing, and promotion. However, the markups, that is, the increases in the price of the item that can be derived from advertising, all depend on the direct costs of production. Therefore, the fabric costs define the final garment value, because fabric offers the features of finery, quality, and originality that allow *upmarketing* of particular garment features.

However, weaving cotton into fabric is also the most ecologically damaging industrial production link because of the chemical pollutants expelled in the liquid effluents that result from the runoff processes during textile manufacturing (Ibrahim et al., 2008; Khan et al., 2009; Pan et al., 2008). In addition, the ecological damage from chemical pollution adds to a fairly large global carbon footprint of high direct fuel costs and their indirect pollution impact in the form of CO_2 emissions. Emissions are significant because the largest cotton exporter in the world is the United States. However, America does not really export its cotton, but, rather, temporarily moves it overseas for processing. In actuality, over 80%

of exported raw US cotton ends up returning to the United States in the form of ready-made garments (Birnbaum, 2008).

Such a platform is based on long supply chains of shipping and trucking cotton, apparel components, and finished garments around the world, in which the average cotton bale is grown in the United States, processed into fabric in India, sewn into garments in China, and then reimported into the United States for final sale. MacDonald et al. (2010) show the relatively high volume of cotton-based imports in relation to overall textile imports into the United States. The problem is that these cotton-based imports are not really imports, because they are fabrics woven from American-grown cotton that has been shipped to overseas textile mills for spinning and weaving. This is evidence of the outcomes the beneficial preferential rules-of-origin tariff exemptions of the US Import Code can create in the form of perverse incentives.

Perverse incentives are related to unintended consequences. They occur when negative, and often unforeseen, consequences form that are contrary to the intent of the incentive provider. Perverse incentives in industrial production are not always unforeseen. In fact, they are often well anticipated and accounted for, but their impact is discounted for various reasons. Such is the case in the apparel trade. As discussed in Chapters 3 and 4, American apparel businesses—from cotton farmers, to textile and garment manufacturers, to retailers—have spent decades asking for market-protecting incentives from the US government. These demands have varied from policies of trade restriction to trade liberalization, and often have had conflicting features. The end result of this legacy of asking the federal government to intervene in the dynamics of the international fashion market is the *rules-of-origin* tariff exemptions of the US Import Code. Under these exemptions, most components that are made from US inputs are given tariff- and duty-free entry into the United States. The customs documentation must provide evidence that at least 40% of the value of the good can be attributed to US inputs. In garment imports, most of this value is documented as coming from American cotton.

Turning a bale of cotton into fabric is investment intensive, energy intensive, and environmentally taxing (Banuri, 1998; Diebäcker, 2000; Greer et al., 2010; Ibrahim et al., 2008; Khan et al., 2009; Tufekci et al., 2007). Textile processing consists of three major industrial operations—pinning, weaving, and finishing. Spinning entails mostly dry processing and generates noise and dust pollution. Spinning is considered the least-impacting stage when it comes to the environment, but it is still very harmful to workers. The average textile plant uses shuttle looms that cause noise levels as high as 100 dB, exceeding the highest safety limit of 85 dB (Pan et al., 2008).

The most serious environmental problems are associated with the wet-finishing processes of fabric manufacturing (Chaturvedi and Nagpal, 2003). The main wet processes are bleaching, mercerizing, and dyeing, which produce liquid effluent with varying waste composition (Banuri, 1998; Tufekci et al., 2007). Before it is woven, cotton needs to be bleached. Then it is mercerized. Mercerizing is the process of dipping the bleached fibers into a sodium hydroxide bath, then neutralizing them in an acid bath (Wakelyn et al., 2006, p. 74). Not all cotton needs to be mercerized, but most of it is, because fabrics treated in this way respond better to dyeing and are used in the production of cotton/polyester blends.

Once cotton is mercerized, it is spun into yarn and then woven into fabric, which is then dyed. During the weaving process, starch is applied to the fabric to impart fiber strength and stiffness, resulting in wastewaters that contain large amounts of industrial-grade starch. During the spinning process, yarns are treated for fineness and texture, usually through wet-heating processes that solidify the starch and apply primary dyes and chemicals that define the fineness and durability of different fabrics that are woven from the yarns. Also in the weaving process, synthetic fabrics can be blended to create poly-blend textiles. The blending process is also heat intensive, resulting in the discharge of large amounts of heated water into local watersheds.

After weaving, fabrics are dyed based on pattern specifications. More starch, sodium hydroxide, and chemical dyes are used during these fabric-finishing processes. The amount and variety of wetting agents, acids, alkalis, and dyes depends on the quality and desired refinement of the end product (Wakelyn et al., 2006). The higher the quality, and subsequently the price of the fabric, the more chemically intensive the manufacturing process.

The wet processes are the most significant components of production, demanding large quantities of water for the different steps in dyeing and finishing, as well as the use of quality petrochemical products. Often, the methods employed are inefficient. Over 15% of the world's total production of dyes is lost during the dyeing of fabrics (Ibrahim, 2008). On average, 200 tons of water are used for every ton of textile produced (Greer et al., 2010). The used water is chemically laden and poses serious environmental threats. The chemical compounds, metals, and toxic substances must be discharged in a runoff process. They travel from the waters around textile plants into the groundwater systems of large regions, affecting the toxicity of entire ecosystems (Ibrahim et al., 2008).

The World Bank identifies 72 toxic elements that are emitted during textile manufacturing. Thirty of them cannot be treated in a purification process. Only 42 of the 72 toxic elements can go through what has become a partial purification process whereby their levels are minimized but not eliminated (Kant, 2012). The World Trade Organization (WTO) sets guidelines for discharge levels, but each country is free to establish its own regulatory structure, determine tolerable discharge levels, and implement oversight and enforcement measures (Chaturvedi and Nagpal, 2003).

WTO water quality standards are classified into aggregate measures and maximum allowable concentrations of specific chemicals. The aggregate measures are pH value (which determines acidity or alkalinity of the effluent), temperature, biological oxygen demand (BOD), chemical oxygen demand (COD), total suspended solids (TSS) or nonfilterable residue, total dissolved solids (TDS), and color (Banuri, 1998). However, the World Bank does not record or keep track of compliance with these maximum allowable COD, TSS, and TDS levels. It tracks BOD metrics instead, because they are easier to test; only one test is necessary for the estimation of BOD levels. TSS and TDS tests must test for each element one at a time to estimate lead, cadmium, or mercury levels. Such testing is expensive, and therefore the WTO delegates it to the individual governments to conduct. Compliance with WTO guidelines is a condition for membership, but noncompliance is not a condition for expulsion. When nations are found to be in violation of guidelines, there is no punitive mechanism vested in the organization. The path for recourse is for a plaintiff nation to file an official complaint against an infracting nation, offering evidence that the noncompliance policies in question have resulted in significant trade distortions. There is no evidence that any such claim has been filed with the WTO based on complaints of environmental noncompliance.

The environmental laws set by the WTO as a condition for membership do not seem to have any effect. Factories generally do not comply with them (Tufekci, 2007; Chaturvedi and Nagpal, 2003; Khan et al., 2009; Pan et al., 2008). They consider investing in treatment technology a nonproductive use of funds in an industry that struggles against strong cost pressures. Treatment is regularly below standards and is rarely checked by the factory, environmental department, or buyer (Khan et al., 2009, p. 66). In China, for example, standards vary across regions because of centrally planned development policies. Many local governments allow companies to emit waste beyond legal limits (Pan et al., 2008).

As clothing sales are growing in both emerging and developed markets, activists are concerned with the increasing environmental and sustainability problems in the production process. Growth in industrialization in the developing world has increased demand for cotton for both domestic consumption and export dependence. This demand has created incentives to support cotton production and exports through policies that allow prices to remain low.

King Cotton: The Economic Power of Cotton Producers

Prior to 2001, US cotton exports were generally in the range of 5–7 million bales per year. In 2005, US exports reached 18 million bales, as quoted by the US Department of Agriculture (USDA) (Abdelnour, 2007). American export volumes further doubled between 2009 and 2010, rising to record highs due to strong demand, mostly from China (Patton, 2010). According to USDA, as of October 2014, US cotton exports have grown even more for China, South Korea, Turkey, Indonesia, and Peru.*

Several researchers have explored the problems of subsidizing cotton farming in America and the issues these subsidies raise for free trade (Heinisch, 2006; Watkins, 2002; Minot and Daniels, 2005; Pan et al., 2010). The charge is that American cotton subsidies distort trade patterns and keep poor nations from increasing their cotton exports, contributing to poverty in the developing world (Abdelnour and Paterson, 2007; Sumner, 2003, 2006, 2007). The WTO Doha development round, which started in 2001, includes the Cotton Initiative. Its inclusion was championed by the activism of four African nations—Benin, Burkina Faso, Chad, and Mali (Anderson and Valenzuela, 2007). The mission of the Doha Development Agenda (DDA) is to facilitate agricultural trade through gradual decrease and eventual elimination of subsidies and tariffs. Because of this mission, the scope of the round is augmented to also include a provision of improving efficiency in agricultural productivity, because most of the nations heavily reliant on agricultural exports are fairly poor African and Asian nations. In the round, cotton trades and production take a major focus (Sumner, 2006). Anderson and Valenzuela (2007) explain that the focus

---

* Weekly change in export volumes is available at http://www.fas.usda.gov/export-sales/cottfax.htm.

on cotton is a result of the victory against the US government in WTO litigation for subsidizing its cotton producers to levels that reach trade distortion magnitude. The Cotton Initiative has also gained support because it has a development component. It argues for the adoption of genetically modified (GM) cotton proliferation in lesser-developed nations. GM cotton is cheaper to produce and would allow these nations to export at prices comparable to the lower prices the highly efficient (and subsidized) American exporters can offer.

LDCs are heavily reliant on cotton production and have suffered losses because world cotton prices have been gradually decreasing since the mid-1990s (Borders and Burnett, 2006). Although global demand for cotton has increased, prices have fallen. It can be argued that the falling prices are mostly due to the subsidization of American farmers. Although lower prices should result in lower production, the United States has doubled cotton production over this same period of time (Borders and Burnett, 2006, p. 1). Minot and Daniles (2005) argue that US cotton subsidies have greatly harmed West African growers by both obstructing access to the US market and also artificially lowering world market prices. In 2001–2002, America's 25,000 cotton farmers received a $230 subsidy for every acre of cotton planted—a total of $3.9 billion. By comparison, wheat and maize subsidies are around $40–$50 per acre. Because of these subsidies, American cotton farmers receive up to 73% more than the world market price for their cotton crop (Borders and Burnett, 2006). During that same period, American cotton merchants "dumped" up to 11 million bales of cotton on the world market at prices as low as 28 cents a pound in 2000, equalizing at about 48 cents a pound in 2004, as reported by the Organic Consumers Association (OCA), a nonprofit consumer protection group.* There are 480 pounds in a bale of raw cotton. Dumping is selling at prices lower than the cost of production. OCA states that at the time, selling cotton at 48 cents a pound meant that it was sold at about 30 cents less than the cost of growing it and at 40 cents less than storing and transporting it to cotton gins. After harvesting, cotton is *ginned*, whereby the leaves, fibers, and twigs, also referred to as *gin trash*, are removed, and seeds are separated and sold separately for cottonseed oil. Farmers deliver the picked cotton to gins, and the gins in turn process it into bales and sell it to cotton merchants. Ginning cotton is its own commercial process, because cottonseed oil and gin trash are sold to the food and livestock

---

* http://www.organicconsumers.org.

industries. Free of gin trash and cotton seed, the cotton bales are sold to cotton merchants, who transport, warehouse, and deliver the bales to the domestic mills or to the shipping yards, where transnational transportation firms handle the final delivery to international buyers. In many cases, these buyers are commodity brokers, who in turn resell the bales to individual mills in different countries, where eventually the bales are spun into fibers.

In this process, the transportation and warehousing costs, also called *bases*, double the price of cotton, and their levels are essential in determining the price of exported US cotton. Because of their fluctuations, merchants trade in cotton futures on the Chicago Board of Trade. They purchase *futures* at a locked price months in advance in an attempt to insure against possible price fluctuations. At the date of delivery, if the futures price is below the market price that day, the cotton merchant makes a profit when the mill "accepts delivery."

In the United States, the lower the price of cotton at the gin, the higher the profit margin for cotton merchants, because their prices are compared with those of other cotton-exporting nations. Because most of these exporting nations do not have the productivity of American farmers, or the price-support programs the American government offers its farmers, their prices are higher and less elastic. The price-support programs the American cotton industry enjoys are direct payments to cotton farmers; crop insurance subsidy, which is a direct payment for the compensation of crop insurance also payable to the farmers; and the Export Credit Guarantee Program, which extends credit guarantees to private US banks for accepting loans from approved foreign banks that represent foreign merchants. It also provides marketing loans at favorable rates to merchants and brokers, and countercyclical payments, available to farmers and ginners, which are an extra insurance against low crop prices. It is an insurance against competition. If prices fall below levels of futures purchased months in advance, then the US government steps in to make up the difference of the loss.

The countercyclical payment program was implemented in 2002—the same year that Brazil, Australia, and the West and Central African (WCA) countries won a complaint with the WTO against the US. The WTO sent its concurrence with the complaint to its Dispute Settlement Body (DSB), which approached the US Office of Trade Representatives (USTR), the legal agency that represents the United States in front of the WTO, to work out the terms of action. The USTR filed an official appeal. The United States offered evidence of price-support activities in the plaintiff nations.

Many nations engage in price-support policies, especially nations with more centrally planned economies, which is the case with the WCA nations that joined Brazil and Australia. For example, WCA cotton production is owned by the individual state-owned cotton companies, which all have a joint-venture partnership with a French state-owned conglomerate, Compagnie Française de Développement des Fibres (CFDT) (Baffes, 2007). This arrangement created a monopsony in cotton buying and a monopoly in cotton processing and marketing, which are both against WTO private rule directives (Baffes, 2011; Varella, 2014).

In the case of the WCA nations, private rule directives can be circumvented because under Article 9 of the WTO, certain underdeveloped nations are allowed to support monopoly and monopsony structures in markets where development levels do not allow competitors to emerge. Therefore, their governments own the entities that define the markets. However, if these governments are reliant on external governmental subsidies, as is the case with the French-based CFDT, then it is the French government that benefits from the price-support policies in WCA nations. This is a violation of the WTO privacy directives, because their monopoly and monopsony exemptions do not apply to developed nations. Baffes (2009) explains that typically, CFDT would announce a base price even before cotton planting, and would often supplement that price with second payments.

Due to such legitimate complaints from the United States, it took 2 years of investigation and negotiation for the WTO to issue its final ruling. On September 8, 2004, the WTO ruled that the United States had to "remove" the adverse effects of the subsidies or "withdraw" the subsidies (Baffes, 2010). The ruling, in effect, said that the United States had two choices—to cease its price-support programs or to pay penalties to the plaintiff nations in the amount of price distortion the subsidies caused. Of course, once again, the United States appealed, but the Appellate Body upheld the original ruling (Baffe, 2011).

The WTO awarded Brazil $830 million, an amount much lower than the requested $3 billion. This was the first time an actual monetary settlement was reached in direct compensation at the WTO. Further, the WTO allowed Brazil to impose countermeasures on American imports from industrial sectors of its choosing, including sectors outside merchandise trade, such as intellectual property and services (Baffes, 2010). The countermeasures were to be allowed starting in 2010. It is interesting that the provision to impose countermeasures was not issued until 8 years after the original ruling of trade distortion. Countermeasures are usually among the first benefits plaintiff nations are awarded (Epstein et al., 2009). This is the case because the WTO does not have a precedent rule, and each case is adjudicated on a case-by-case basis (Zweifel, 2006). Through awarding countermeasures, the WTO just sends the two arguing nations back to the negotiating table, telling them which one of them is going to be the winner. The winning nation gets a financial settlement from the losing nation. The amount of this settlement, however, has to be agreed on upfront; if it is not agreed on, it is understood that the disputing nations cannot come back to the WTO with further complaints. Therefore, nations engage in short-term countermeasures, usually until the injured party is able to make up some portion of its losses, and then the measures are lifted.

But Brazil is not just any nation; it is a major BRIC trading partner of the United States. Noncotton US MNCs are expanding rapidly in its rich oil fields, dominating its energy sector, and benefitting from its vast manufacturing capabilities for both export and manufacturing for the rapidly growing purchasing power of the local population. Clearly, it is no coincidence that countermeasures were stalled, suggesting that the US position was to strongly object to them.

It is clear that the demands of the plaintiff nations were not met with respect to the ceasing or easing of US government price-support policies for cotton exporters. According to the Environmental Working Group, a nonprofit advocacy think tank headquartered in Washington, DC,* total cotton subsidy amounts have remained unchanged, averaging over $1 billion a year since the 2004 ruling, and rising as high as $3.6 billion in 2005. The data also reveal that between 2005 and 2009, most of the amounts were in direct payments and production flexibility contracts, roughly equally matched by marketing loans and loan-deficiency direct support, where crop insurance premium subsidies were the smallest portion of price-support compensations. Starting in 2010, however, crop insurance premiums overtook all other forms of assistance, with no amounts shown for countercyclical programs, direct payments, and production flexibility contracts.

---

* Data available at farm.ewg.org/.

In April 2010, the United States and Brazil signed a memorandum of understanding (MOU) for "paving the way to settle the dispute" (Baffes, 2011, p. 1534). It can be concluded that the disappearance of direct payments and payments through countercyclical programs can be attributed to concessions by the United States in "paving the way" toward settlement, as Baffes (2011) puts it. However, since the total amounts of support have not diminished significantly—only its form has change from direct payments to subsidizing insurance premiums—it can be concluded that things are proceeding in a business-as-usual fashion. If anything, doing business the "American way"—through subsidizing production—may be proliferating: interestingly, the 2010 MOU included, shockingly, a new subsidy, this time payable to Brazil. It is ironic that the very concept of the complaint—subsidizing—was used as an appeasement offering in the complaint resolution. Of course, the subsidy was not called a subsidy to the Brazilian government. It was called a "fund" in the form of an annual payment of $148 million to aid "capacity building" for the Brazilian cotton industry (Baffes, 2011). The fund was to remain until the next US farm bill was passed by Congress in 2012, and was signed into law by President Obama, effective February 7, 2014. As soon as the Farm Bill passed both houses of Congress, the United States eased off on the payments. In the first half of 2013, the United States decreased the amount of the payment and failed to complete it at the end of fiscal 2013. As a result, Brazil filed a new complaint with the WTO. The WTO allowed Brazil to "request" a new panel to investigate whether the new farm bill would bring the United States into compliance with previous rulings. The Brazilian government decided not to request the panel, but to enter into bilateral negotiations with the United States (Meyer, 2014).

This case is an example of what is wrong with the current system of international trade compliance and adjudication of disputes. In the most basic of crops, the most vital of commodities, the largest nonfood traded good on earth, so many parties that have a shared interest in efficiently producing cotton cannot agree on what constitutes fair trade in terms of market distortion. The example also illustrates the symbolic role of the WTO and its inability to litigate swiftly and decisively against the actions of member governments, particularly against the action of the government of the nation which is its main funder—the United States.

The case was a first for the WTO, because it was filed jointly, it awarded cash compensation, and its adjudication has taken a decade. At the initial ruling in 2004, environmentalists praised it (Pan et al., 2010). The environmental implications were that reducing the price support that was keeping the world price of cotton artificially low would allow relatively more expensive cotton, such as organic cotton, to be more competitive. The expectation was that retailers then would have stronger incentives to introduce organic cotton lines.

Many environmentalists praise organically grown cotton because of its lower use of pesticide (Dem et al., 2007). But organically grown cotton has raised objections from human rights activists because it is much more water intensive. In developing nations, populations must choose between using precious water for organic cotton farming or for daily human necessities such as proper hydration and hygiene (Eyhorn, 2007). Furthermore, as already explained, dyeing the fibers is much more pollution intensive than pesticide use. This fact turns organic cotton into a very "inorganic" fabric. It is all very clearly explained and justified by Nimon and Beghin (1999), who find that while customers are willing to pay a premium for organic cotton, they are not willing to pay the premiums for environmentally friendly dyes. Paying both premiums puts the garment out of desired price ranges.

Organic or not, farming is only one link in the long supply chains of cotton products. According to a 2007/2008 study on the growth of organic cotton farming globally, farm-level costs are a very small part of the total cost in the supply chain of garment manufacturing (Feriggno, 2009). Still, organic cotton is popular. Ellis et al. (2012) examine customer willingness to pay premiums for garments made from organic cotton, and in their study the surveyed participants, on average, displayed willingness to pay up to a 25% premium. However, only those participants who did not have financial constraints were willing to pay such a premium. Participants who pay for their own clothing, that is, older, wage earners, who are the primary decision makers for their own clothing expenditures, were not willing to pay the premium. Interestingly, just as found in previous studies on willingness to pay organic premiums (Nimon and Beghin, 1999; Loureiro et al., 2001), willingness to pay is mostly dependent on sociodemographic control factors, such as race and level of income.

Succinctly, Ellis et al. (2012) say that managers need to understand the income levels of their target customers when deciding on organic options. Such recommendations are targeted at managers in the developed world. However, as already discussed, much of the production of fabric and the manufacturing of garments is done in the developing world. Much of the consumption growth also has risen there in recent years. And yet, the growth in the agricultural production of cotton has been the strongest and most concentrated in the United States. One can argue that this fact is based on the American longstanding, complicated, and politically driven infrastructure of providing superior government support to cotton producers. It can also be said that it is the technological aptitude in machinery, fertilization, and irrigation that allows American cotton farmers to increase their productivity. For these reasons, it seems a little less counterintuitive that the US textile industry has gradually declined in relation to its developing nation competitors. Why does American cotton have to be exported to these nations? Pan et al. (2008) argue that the trade distortions in cotton production have contributed to environmental pressures in countries that are most negatively affected by falling cotton prices. They face incentives to develop their internal cotton-processing industries, as opposed to exporting cotton, when cotton prices are too low. Because these nations lack the industrialization platforms to develop their own fiber-processing industries, they compete for foreign investors to develop the sector. In this way, they hope to offer an internal market for the locally produced cotton that cannot compete on the world market because it needs to be priced higher than the American options.

In this quest to attract investors, LDCs try to offer the most attractive production environments. Because cotton processing is highly environmentally taxing, and paying for environmental cleanup in textile production is relatively expensive, LDCs face incentives to lower and/or eliminate their environmental standards, provided they have the institutional structure to have even implemented environmental standards in the first place.

EXPORTING COTTON = EXPORTING POLLUTION

The most often-repeated sustainability concerns center around the carbon footprint of supplying American cotton to the leading global textile-processing centers—China, India, Pakistan, and Vietnam. The United States has strengthened its leadership as the number one exporter of raw cotton. According to the USDA, in 2012 more than 75% of the cotton produced in America was exported. In contrast, less than 40% of American cotton was exported in 1990. And according to Birnbaum (2008), America then accounted for almost 40% of world cotton exports. Today, the OCA argues that the number has risen to close to 60%.

Today, more than 45% of China's cotton imports come from the United States, and that percentage is expected to increase in order to sustain the growth of the Chinese textile industry. The Chinese government's official policy of managing the country's cotton-processing needs is to increase imports, since China does not have a large-scale agricultural production. According to Pan et al. (2008), because of limited agricultural capacity, priority is given to growing food crops for China's growing food consumption rather than growing cotton. The state's role in managing the cotton industry is very strong. In 2008, more than 33% of China's imported cotton went to state-run enterprises.

Analyzing production location trends in the industry, the United Nations Conference on Trade and Development issued a comprehensive study edited by Miroux and Sauvant (2005) identifying the leading global nations whose economies are dependent on apparel production. The report made special distinction between the different classifications of apparel. In Pakistan, one of the leading exporters of both textiles and apparel, textiles have grown to comprise over half of all merchandise exports. In India, apparel exports account for 55% of all export earnings. However, only about 12% of these exports are in the form of ready-made garments, so that 88% of exports classified under "apparel" are actually in the form of fabric (Chaturvedi and Nagpal, 2003). The other global leaders in textile exports are Nepal (16%), Macao (China) 12%, Turkey (11%), and India (11%) (Miroux and Sauvant, 2005, p. 4).

The distinction is very important, because most research has not made a division between textiles and apparel. Researchers continue to group them together when studying exports (Miroux and Sauvant, 2005, p. 4). The main reasons are historic and academic. Most industrialization literature has developed around the flying geese model. This model was introduced by Japanese economist Akamatsu, who studied German manufacturing and FDI in the 1930s with a particular focus on its links to the Japanese textile industry (Akamatsu, 1961, 1962). Akamatsu examined how textile mills helped develop the local chemical industry by fostering local partnerships in order to develop cheaper alternatives to shipping dyes from Germany. As the Japanese chemical industry grew from its German FDI roots, it attracted other investors who helped develop pharmaceutical-industrial platforms and the petrochemical refinery industry, and these in turn attracted more investors, who gradually built a diverse Japanese manufacturing base. It all started with the "leading goose"—the textile industry. This legacy has led to the argument that historically, textiles and clothing have been the foundation for industrialization and development of many countries (Gereffi, 1999).

Researchers on industrial upgrading, the overall development of nations from agrarian and subsistence agriculture dependence to manufacture-driven economies based on exports, continue to apply the model and its core assumption. The core assumption is that the apparel industry as a whole is often the lead goose that starts the industrial upgrading process (Akamatsu, 1961, 1962; Adhikari and Yamamoto, 2007; Brautignam, 2008; Kumagai, 2008; Schroeppel and Nakajima, 2002). The policy implication of this assumption is that the development of the apparel sector should be supported because it would lead to industrial growth in other sectors. The problem is that the MFA changed that assumption, and progress in manufacturing, merchandising, and retail management has created clear divisions between the different industrial processes in apparel production. In the first place, the flying geese model is based on the assumption that foreign investors have a choice in location. They respond to holes in the local market and direct assets in industrial development in nations that show a need for the development of certain sectors. But in the case of all apparel—textile, garment assembly, and trim production—the quota system put significant location restrictions on garment producers in 1960. Second, as already discussed, textile manufacturing today is very different from in Akamatsu's day, as technology has evolved, and textiles have changed from simpler fabrics of cotton and wool to a complicated array of poly-blend materials. Foreign direct investment in the two sectors has different purposes and needs. While textile production has cost-minimizing needs in inputs, such as petrochemical products and utilities, garment production has cost-minimizing needs in labor. Textile production is contingent on the proximity and availability of dyestuffs, solvents, and energy, implying that reasonable development in these industries would be required, as well as the presence of an adequate transportation infrastructure for speedy movement of large amounts of both inputs and outputs.

As economic theory states, any economic activity increases efficiency when the degree of specialization grows. This fact holds true for garment production, as the evidence suggests that the industry has developed clearly specialized production niches. It is also true that this specialization is evident at the country level. When it comes to exports, certain countries, such as the United States, specialize in the agricultural link of the production chain—the growing and selling of cotton. Certain nations, such as Hong Kong, specialize in final assembly and shipping of finished garments.

Certain nations, such as China, strive to be a total component specialist. However, for most nations, as Miroux and Sauvant (2005) show, it is textile production as a specialization that defines exports. The cotton growers are not the leading exports of textiles, and the final assembly manufacturers do not have significant local textile production. The leading exporters of textiles, as identified by the United Nations, are listed in Table 5.1.

Among the 33 nations, the majority are fairly small countries. The Maldives, Fiji, and Northern Mariana Islands have a population of less than 500,000. Mauritius is barely 1.5 million, and more than half of all nations have populations less than 10 million, such as Bulgaria, Macedonia, Latvia, Lithuania, Estonia, Cape Verde, Haiti, Lao, and Tunisia. With the exception of South Korea, all the top leading global exporters of textiles are fairly poor developing nations, with many being classified as very under-developed, such as Cambodia, Lao DPR, Cape Verde, Haiti, Madagascar, Lesotho, Nepal, Pakistan, and Sri Lanka. Tunisia is so low on international developmental metrics that its population estimates come with the disclaimer that because over 98% of its citizens lead a nomadic existence in desert migration, it is very hard to estimate any sociodemographic metrics that can gage population welfare.

It is counterintuitive that such small nations, in terms of both population and geography, can supply large enough quantities of textiles to be ranked next to the large export capacities of China and India, with populations in the billions, Pakistan and Indonesia, with populations in the hundreds of millions, South Korea, with over 50 million, and Turkey, with 76 million. It is also counterintuitive, and very telling, that those tiny nations have negligible agricultural capacities. The implications are that they do not grow their own cotton and other natural fibers, such as linen, or raise sheep for the production of wool in quantities large enough to sustain an integrated local textile industry. Just from that list alone, one can deduce that they import their textile inputs.

Based on the numerous examples from previous research that in developing nations the textile production facilities are owned by foreign MNCs, the question arises: why do the MNCs locate in those particular nations? Few reasons can be deduced at first glance. One reason is proximity to major markets. The presence of many East European and African nations suggests that being close to the largest market in the world—that of the European Union (EU)—is important. Product from Eastern Europe can be trucked to major EU distribution centers in 48 hours. Product from Tunisia, Turkey, and Egypt can be shipped to European ports in as little

TABLE 5.1

Leading Exporters of Textiles

Country	Region	Net FDI in Million US$		H_2O Pollution from Textiles		H_2O Pollution from Chemicals		Apparel as % of Manufacturing	
Year		1991	2008	1991	2008	1991	2008	1991	2008
Albania	Europe	20.00	1.00	59.80	60.19	—	—	26.5	22.30
Bangladesh	Asia	1.39	973.11	77.11	77.11	3.22	3.22	42.2	42.2
Belarus	Europe	—	21.49	—	—	—	—	—	—
Bulgaria	Europe	55.90	8,472.19	20.68	28.04	10.52	10.52	11.7	13.44
Cambodia	Southeast Asia	33.00	794.69	6.83	59.35	33.51	33.51	2.6	86.5
Cape Verde	Africa	1.20	213.83	—	—	—	—	—	—
China	Southeast Asia	3,453.00	94,320.09	—	—	—	—	1.67	2.2
Czech Republic	Europe	564.36	8,966.89	15.21	7.40	7.08	10.89	5.6	3.35
Dominican Republic	Latin America	145.00	2,884.70	73.07	73.07	2.34	2.34	—	—
Egypt, Arab Republic	Africa	191.00	7,574.40	31.11	31.11	13.88	13.88	13.23	10.06
Estonia	Europe	80.40	875.93	23.62	8.78	6.72	8.42	15.01	4.59
Fiji	Southeast Asia	11.93	332.67	38.56	38.56	4.25	4.25	20.13	21.77
Haiti	Latin America	11.80	29.80	0.00	—	0.00	—	—	—
Hong Kong	Southeast Asia	—	3,082.98	—	—	—	—	—	—

Country	Region								
India	Asia	73.54	22,807.03	—	—	—	12.77	—	11.75	8.68
Indonesia	Southeast Asia	1,482.00	3,418.72	31.61	31.61	12.77	12.77	17.55	11.53
Korea, Republic	Asia	−308.80	−10,594.70	24.99	9.34	9.62	12.05	13.31	4.94
Lao PDR	Southeast Asia	6.90	—	—	—	—	—	21.97	21.99
Latvia	Europe	27.29	1,092.00	19.93	12.61	5.61	5.59	11.94	7.1
Lesotho	Africa	273.59	218.04	90.14	90.75	0.79	1.20	—	—
Lithuania	Europe	30.18	1,383.37	23.30	19.33	5.66	7.57	17.99	10.88
Macao	Southeast Asia	—	3,494.25	—	—	—	—	—	—
Macedonia	Europe	—	612.03	—	—	—	—	16.88	16.69
Madagascar	Africa	13.68	85.44	59.93	58.95	11.71	12.38	35.26	30.93
Maldives	Southeast Asia	6.50	15.43	—	—	—	—	—	—
Mauritius	Africa	6.52	325.30	—	—	—	—	51.98	42.44
Nepal	Asia	19.16	1.00	38.66	38.66	5.78	5.78	33.54	19.26
Northern Mariana Islands	Southeast Asia	—	—	—	—	—	—	—	—
Pakistan	Asia	262.15	5,389.00	—	—	—	—	—	—
Sri Lanka	Southeast Asia	43.83	690.50	43.56	43.56	8.96	8.96	29.45	29.45
Tunisia	Africa	122.21	2,600.67	—	—	—	—	—	—
Turkey	Europe	783.00	15,414.00	30.27	35.66	8.34	9.77	15.67	17.59

as 24 hours. Product from the rest of the African nations on the list can reach Europe in as little as a week. For these reasons, there has also been a noted investment in Africa from Chinese and Asian textile MNCs. Azmeh and Nadvi (2013) explain how Chinese investments are building the textile sector in Jordan. Proximity to the EU market is noted as a core reason for the investment.

There is also a specific geographic reason for the larger Asian exporters to be on the list. They are all close to China, or have historic legacies of abundance in natural endowments, such as land and labor, that made them important textile and garment industrial centers during the quota years. Such is the case with Pakistan and Bangladesh, as well as Turkey, where the cotton trade and textile production go back centuries. Proximity to mainland China allows speedy integrated production for Chinese export, and as Birnbaum (2008) and Lu (2012) explain, Chinese factories have been able to compete through offering such complete services as being able to deliver a finished garment from a simple sketch, undertaking all links in the manufacturing process, so that the only thing a client needs to do is write a check. The other reason why proximity to China is important is its own growing market. As already discussed, particularly with respect to fast-fashion retailers, the largest number of new store openings has been in mainland China.

However, proximity is a fairly recent advantage in location because during the MFA years, proximity to market was secondary to the availability of quota. Also, until the recent rise in fuel costs as a result of several factors, chief among which are the instability and war in the Middle East as well as growing energy demand from the developing world, transportation and communication costs had been in a historical decline.

Thomas Freidman (2005) famously said: "the world is flat," meaning that proximity to market was not important anymore, because transportation costs were so low. Given that information, and taking into consideration the volumes of research that outline the main reasons for international site selection of MNC affiliates as (1) local factor endowments, (2) size of local market, (3) growth of the local economy, (4) gaps in the local market, (5) size and growth rate of the local economy, and (6) local strategic benefits, it is even less counterintuitive that these tiny, poor nations attract textile MNCs (Baltagi et al., 2008; Buckley and Ghauri, 2004; Chowdhury and Mavrotas, 2006; Kinda, 2010; Porter, 2000; Zhang et al., 2014). None of the nations have factor endowments that are conducive to large-scale textile production. Their local apparel markets are extremely small. Their local economies may be growing, but for all these nations except South Korea, the growth rates are far below the necessary levels to elicit local-market growth. The poverty in those nations is such that there are gaps in the local market, that is for sure, but the size of the local market is too small to require significant international investment. Very few of the nations are located in strategic commercial, diplomatic, or military locales. Because none of the six factors for increasing foreign investment is obvious at first, there is strong evidence that elements of environmental sourcing might be at play.

Environmental sourcing, as already discussed, is the location of foreign firms in nations with lax regulatory environments in order to benefit from lower or nonexistent environmental regulation that would allow businesses not to employ costly pollution-mitigating technologies. The World Bank offers approximate metrics on pollution from industrial activity. For textiles, it offers an aggregate BOD pollution measure. As already explained, this is an overall measure of pollution that is not the best, but it is useful to examine, because the World Bank approximates what percentage of the total industrial water effluents come from the textile industry. Table 5.2 includes these numbers for the top textile exporters and the change in these levels from 1995 to 2008. Also included are the BOD pollution levels associated with chemical production, because textile production is dependent on the availability of chemical inputs. The relationship between both metrics is important in production lead times, as short as they are currently in fast-fashion assembly. It is as costly as it is time consuming to be shipping in commercial amounts of peroxide, bleach, and dyestuffs from far-off importers. The other two metrics offered in the context of textile production are total percentage of the local economy that is reliant on apparel manufacturing and yearly foreign direct investment inflows. Here is an example of research grouping textiles and apparel into one category. The World Bank's World Development Indicators database is the only one that offers even that level of granulation. It is impossible to separate the exact breakdown of textile and clothing manufacturing, and apart from the few examples already offered of the difference, consistent metrics per country and per year are not available, even though the overall measure is important and merits analysis. It is also important to examine how foreign investment inflows are influenced by these factors. The main challenge is data availability. As Table 5.1 indicates, many of the nations do not report their pollution levels to the World Bank. Extensive investigation for this project started in 2009, and no reliable metrics could be identified.

TABLE 5.2

Textile Exporters with Available Pollution Metrics

Country	Region	Net FDI in US$ Million		% H₂O Pollution from Textiles		% H₂O Pollution from Chemicals		Apparel as % of Manufacturing	
		1995	2008	1995	2008	1995	2008	1995	2008
Albania	Europe	20.00	843.68	59.80	60.19	—	—	26.50	22.30
Bangladesh	Asia	1.39	973.11	77.11	77.11	3.22	3.22	44.50	42.22
Bulgaria	**Europe**	**55.90**	**8,472.19**	**20.68**	**28.04**	**10.52**	**10.52**	**10.90**	**13.44**
Cambodia	**Southeast Asia**	**33.00**	**794.69**	**6.83**	**59.35**	**33.51**	**33.51**	**22.20**	**86.50**
Czech Republic	Europe	564.36	8,966.89	15.21	7.40	7.08	10.89	7.20	3.35
Dominican Republic	Latin American	145.00	2,884.70	73.07	73.07	2.34	2.34	—	—
Egypt	Africa	191.00	7,574.40	31.11	31.11	13.88	13.88	13.23	10.62
Estonia	Europe	80.40	875.93	23.62	8.78	6.72	8.42	15.01	4.59
Fiji	Southeast Asia	11.93	332.67	38.56	38.56	4.25	4.25	20.12	21.77
Indonesia	Southeast Asia	1,482.00	3,418.72	31.61	31.61	12.77	12.77	17.55	11.53
South Korea	Asia	−308.80	−10,594.70	24.99	9.34	9.62	12.05	9.52	4.94
Latvia	Europe	27.29	1,092.00	19.93	12.61	5.61	5.59	11.94	7.10
Lesotho	Africa	273.59	218.04	90.14	90.75	0.79	1.20	—	—
Lithuania	Europe	30.18	1,383.37	23.30	19.33	5.66	7.57	17.99	10.88
Madagascar	Africa	13.68	85.44	59.93	58.95	11.71	12.38	35.26	30.93
Nepal	Asia	19.16	1.00	38.66	38.66	5.78	5.78	33.54	19.26
Sri Lanka	Southeast Asia	43.83	690.50	43.56	43.56	8.96	8.96	29.45	29.45
Turkey	**Europe**	**783.00**	**15,414.00**	**30.27**	**35.66**	**8.34**	**9.77**	**17.22**	**17.59**

However, even with such difficulties, some data approximations are available, and the countries for which the data are recorded are included in Table 5.2.

The time frame in Table 5.2 is from 1995 to 2008—the last year with available data. It aims to show how the end of the MFA, in terms of the agreed-upon gradual dismantlement of the system in 1995 as well as its final end in 2005, impacted the economies of the leading textile exporters. The data reveal how their overall apparel sector changed in relation to all other manufacturing sectors, how their pollution levels associated with textile production changed, and which nations became relatively more attractive to foreign investors. The results show that only three nations became relatively more reliant on apparel production—Bulgaria, Turkey, and Cambodia. Cambodia became somewhat of a specialist, dedicating the majority of its manufacturing capacity to apparel. In all other nations, the sector's importance declined. The nations where that decline was strongest are italicized. It is hard to say whether the end of the quota system hurt their textile exports or whether their economies diversified with respect to other manufacturing activity, so that a lower percentage of all manufacturing is attributed to apparel. This seems to be the case with the East European nations that became members of the EU with the accession rounds of 2001. In these nations—Estonia, Latvia, and Lithuania—FDI increased significantly while the apparel industry's importance declined. In the Czech Republic, its importance decreased by half. In these nations, pollution levels also decreased drastically. A possible explanation could be increased incentives to comply with EU regulations. However, in many of the other nations, pollution levels increased, and a notable fluctuation is evident, based on region, reliance on apparel exports, and the ability to attract foreign investment. To examine this variability, three cross-sectional time-series regression analyses provide supporting evidence for the pollution haven hypothesis with respect to textile production. It seems that the end of the quota system increased the incentives for environmental sourcing. Among the 33 leading textile exporters, those nations that allowed pollution levels to rise were able to attract relatively more foreign business.

Using FDI net inflows as a dependent variable, Table 5.3 illustrates that with the end of the MFA, FDI increased in those textile-exporting nations that allowed pollution levels from textile manufacturing to increase.

The data come from the World Bank. They are analyzed with STATA software. As per Greene (2008) chapter 8, a Houseman test is employed to

TABLE 5.3

Influence of Pollution from Apparel Manufacturing 1995–2008, FDI Net Inflows

Variables	Coefficient	Standard Error	Significance
Textile Industry$_{\text{WATER POLLUTION}it-1}$	213.13	96.30	*
Chemical Industry$_{\text{WATER POLLUTION}it-1}$	302.94	225.04	^
Apparel as % of Total Manufacturing	−108.69	74.41	^
WTO Membership	−509.24	434.15	NS
Country			
Bangladesh	10.95	3,724.06	NS
Bulgaria	8,404.81	3,970.61	*
Czech Republic	13,300.22	4,885.51	**
Egypt	6,008.05	3,000.74	*
Estonia	8,130.30	4,949.59	^
Fiji	5,390.84	4,510.70	NS
Indonesia	4,598.61	3,170.89	NS
Latvia	8,738.98	5,414.55	^
Lithuania	7,087.59	4,709.93	^
Madagascar	277.78	2,677.70	NS
Nepal	4,824.99	4,379.78	NS
South Korea	5,444.22	4,177.10	NS
Sri Lanka	4,039.54	3,629.53	NS
Turkey	9,415.65	3,646.86	**
Constant	−10,482.39	7275.57	^
Prob. > F	<.0001		
R-squared	0.40		
Observations	207		

Note: Dependent Variable: FDI net inflows—the overall balance of foreign assets to liabilities in a country measured in millions of current US dollars in a given year.
Fixed effects two-tailed tests.
NS = not significant.
$^{\wedge}p < .10$; $^{*}p < 0.05$; $^{**}p < .01$; $^{***}p < .001$.

examine whether the panels, constructed of individual nations in a given year, show significant degrees of autocorrelation. This autocorrelation could result from similarity of data points in a given country from year to year. The Houseman test revealed that such a correlation could be a problem, and to control for it, a fixed effects test is performed. Apart from offering a way to control for autocorrelation, and therefore provide more reliable estimates of causal relationships, a fixed effects test also allows observation of the differences among the individual countries during the specific years of the time series. In addition, the pollution-independent variables are lagged by 1 year to control for endogeneity, which could be attributed to reverse causality. The reverse causality could stem from the fact that it might be foreign investment that leads to increase in pollution, because the new facilities built with that investment increase industrial output and, therefore, pollution. However, because FDI is not separated into industrial sectors, endogeneity is less likely. The logic is that FDI net inflows come in many forms, including financial instruments, M&As, and even shifting debt for the purpose of minimizing tax liabilities. An increase in FDI does not necessarily indicate an increase in industrial output. A good example in the data is South Korea, which has a negative FDI balance. This fact means that South Korea invests overseas in amounts larger than the foreign investment it receives. But South Korea is a uniquely rapidly developing nation, which moved from developing to developed status during the examined time frame. South Korean MNCs such as Daewoo, Samsung, LG and so forth are among the largest MNCs in the world that invest aggressively outside their borders. Other industrialized nations, such as the Netherlands and Austria, are also negative net foreign investors, so having a negative FDI balance is not uncommon. However, for the other nations in the study, one could argue that because they are so underdeveloped, FDI inflows are more likely to increase because they are essential for economic growth. Furthermore, because of their size and the fact that textile exports are claimed to be the main industrial output of these nations, it is reasonable to deduce that a significant portion of FDI would be directed toward the textile sector. However, looking at the raw data of percentage of local manufacturing associated with the whole apparel sector—including all aspects of garment production—reveals significant variability among the examined nations. Therefore, some reverse causality could be present, particularly with the effluents from manufacturing chemicals. For that reason, pollution levels from both textile and chemical manufacturing are lagged by one time period. Lagging allows the observation of patterns of causality in different periods in such a way that the amount of the dependent variable in a given year is regressed against the independent variables in the following year. In this way, one cannot argue that the value of y has caused the change in the value of x, because x has not yet occurred. Lagging independent variables is particularly useful when examining relatively long time frames in which the pattern of similarity can be evaluated repeatedly.

The results suggest that foreign investors preferred nations where the pollution effluents from textile manufacturing were relatively high and had a pattern of increase through the years. If rising pollution levels are seen as signs of decreasing environmental regulatory rigor, then nations can increase their FDI attractiveness by continuously lowering their environmental standards. Prior studies, such as Shandra et al. (2008), on export support structural adjustment programs in developing nations argue that this is indeed the reality in the majority of poor countries that are competing aggressively to attract foreign businesses. Many of them do so by lowering environmental standards that were nebulous to begin with. The results of the examined textile exporters provide support for these arguments. Both metrics of pollution show statistical significance. The relationship is stronger in the case of textile effluents, indicating that the ability to pollute water basins during the manufacturing of textiles was important for foreign investors. The fact that rising pollution from textile manufacturing was also important has two implications. One is that the industrial connection between the two sectors is very close—textile manufacturing benefits from the availability of local chemicals. The other interpretation is that since chemical production is the core of much industrial production, from textile dyes to pharmaceuticals, to metallurgy, to agriculture, the growth of the chemical industry signals the growth of economic diversification. This second interpretation is supported by the statistically significant relationship between FDI and apparel production as a percentage of all manufacturing. The relationship is negative, suggesting that FDI increased in those countries that gradually became less reliant on apparel manufacturing but diversified in other industrial production. An additional interpretation with respect to textiles is that, as already discussed, textile production has become very capital intensive with significant needs for energy, infrastructure, and financial flexibility. Therefore, investors do not just prefer relatively poor nations; they prefer those nations that show some degree of industrial progress. The implication is that they seek out nations that are poor in terms of regulatory institutions, but not that poor in terms of industrial growth.

The list of nations included in Table 5.2 indicates that Bulgaria, the Czech Republic, and Turkey were the most attractive of the examined nations, in terms of the above-discussed factors. Miroux and Sauvant

(2005) offer Bulgaria as an example. Between 2002 and 2004, it was second only to China in large textile plant construction, with 14 new facilities. During that time, four new integrated manufacturing projects were also constructed, bringing the total apparel infrastructure to 18 new facilities. In comparison, 13 facilities were constructed in nearby Hungary, seven in Poland, and six each in Slovakia, Croatia, and Russia.

With respect to industrial recruitment's ecological impacts in Bulgaria, Dimitrova et al. (1998) examine industrial and managerial behavior that has led to significant and dangerous levels of heavy-metal pollution of the entire watershed of the Devnya region of Bulgaria. Many of the new textile factories built in Bulgaria 5 years after that study were indeed in the Devnya region (Bulgaria Fact Sheet, 2004). The leading investor was Miroglio, SpA from Italy, one of the largest European textile and apparel groups. In the other East European nations of the model, the combined effect of the independent variables also shows statistical significance at the 0.1 level, which is lower in magnitude, but still indicates 90% probability that FDI increased in these nations based on their diversification and rise in pollution levels.

In the context of trade liberalization, the above results provide insight on several points. Between 1995 and 2008, Europe gradually became one union. Eastern Europe was fully integrated between 1995 and 2006, all borders in the shipping of goods were removed, and the implementation of a one-currency system, the euro, was introduced. At the same time, the MFA was gradually removed. Both facts indicate that trade liberalization increased the choice spectrum of textile MNCs, making it easier to locate, should they choose to, in nations that had previously carried higher transaction costs in the site-selection process. Transaction costs are direct and indirect costs of doing business. Low transaction costs indicate ease of doing business; high transaction costs indicate difficulties, which can range from institutional instability to outright prohibition.

Much of the current literature on garment production has focused on the quota system and its removal. To examine how important its dismantlement was in terms of textile MNC site selection, two more regressions are analyzed. Table 5.4 shows the results for FDI inflows based on reliance on the apparel industry and environmental metrics during the 1990s. Although the announcement of the end of the MFA came in 1995, notable policy changes were not implemented until the Doha round starting in 2001 (Gereffi and Memedovich, 2003).

TABLE 5.4

Influence of Pollution from Apparel Manufacturing 1991–1999, FDI Net Inflows

Variables	Coefficient	Standard Error	Significance
Textile Industry$_{\text{WATER POLLUTIONit-1}}$	−19.99	34.84	NS
Chemical Industry$_{\text{WATER POLLUTIONit-1}}$	−12.15	49.48	NS
Apparel as % of Total Manufacturing	9.36	24.37	NS
WTO – WTO Membership	105.04	418.90	NS
Constant	1,578.69	2,119.25	NS
Prob. > F	0.0028		
R-squared	0.38		
Observations	117		

Note: Dependent Variable: FDI net inflows—the overall balance of foreign assets to liabilities in a country measured in millions of current US dollars in a given year.
Fixed effects two-tailed tests.
NS = not significant
$^\wedge p < .10$; $^*p < .05$; $^{**}p < .01$; $^{***}p < .001$.

Neither environmental factors nor the size of the local apparel sector seemed to matter for the increase in foreign investment. This fact shows how strong the prohibitive policies of the MFA were. However, things changed in the 2000s enough to explain the overall trend observed in the data analysis from 1995 to 2008. Table 5.5 shows how FDI increased in the examined nations with respect to the variables of interest.

TABLE 5.5

Influence of Pollution Regression Analyses 2000–2008, FDI Net Inflows

Variables	Coefficient	Standard Error	Significance
Textile Industry$_{\text{WATER POLLUTIONit-1}}$	591.21	199.59	**
Chemical Industry$_{\text{WATER POLLUTIONit-1}}$	995.48	658.76	^
Apparel as % of Total Manufacturing	−119.19	163.23	NS
WTO—WTO Membership	−552.57	555.24	NS
Constant	−10400.75	5558.09	^
Prob. > F	<.0001		
R-squared	0.53		
Observations	132		

Note: Dependent Variable: FDI net inflows—the overall balance of foreign assets to liabilities in a country measured in millions of current US dollars in a given year.
Fixed effects two-tailed tests.
NS = not significant.
$^\wedge p < .10$, $^*p < .05$, $^{**}p < .01$, $^{***}p < .001$.

Rising water pollution levels become not only statistically significant, but also the strongest predictor of FDI increase. Their effect is so strong as to overshadow the size of the local apparel industry as an element of attraction for foreign investors. Also significant, but to a lesser degree, is the pollution from textile manufacturing. The implication is that when the international trade of garments liberalized, environmental sourcing factors become so important for producers as to render direct cost efficiencies to be gained from backward linkages to local industries secondary in choosing where to increase investment.

As already discussed, the data used in the above regressions are based on World Bank estimates and are not the best metrics in terms of gaging actual pollution levels. Based on the case-study analysis of Banuri (1998), Tufeksi et al. (2007), and others, the reality is much worse. These studies are based on data that are not part of the regressions here because they are not available, due to the fact that certain national governments, particularly those of the largest textile and apparel producers—China, India, and Pakistan—choose not to report their overall pollution levels. And yet, there has been much discussion around case-specific evidence of serious environmental damage and its public-health implications in these and other developing nations. Banuri (1998) and Tufeksi et al. (2007) conducted fieldwork in which actual pollution was tested around individual textile mills. In 2008, *National Geographic* published the photo-essay "Pollution in China" by Lu Guang. Guang secretly took pictures of water pollution, documenting the sickness it caused as well as the complacency of the government. One photo shows industrial pipes pumping polluted freshwater deep into the Yellow Sea. The explanation is that instead of investing in cleaning technologies, the official government policy is to save on such costly investments, continue business operations as usual, but try to mitigate their toxic impacts by discharging effluents further from population centers and arable lands, that is, into international waters. It is cheaper to lay the pipe and pump the pollution out than to invest in purification technologies. Many of the examples in Guang's 40-picture essay come from textile and chemical plants. As already mentioned, the majority of textile mills in China at that time were foreign owned (Birnbaum, 2008). As a result, Guang won prestigious awards from Western organizations, and gradually even the disgruntled Chinese government has admitted that water pollution from industrial activity is a serious problem in China.

These facts raise many questions of corporate social responsibility. It is confounding that such a negative reality gradually developed around economic interests based on increasing the production and consumption of inexpensive clothing. It is astounding that as the incentives to keep direct costs low grew with the advent of fast fashion, no industry insiders raised social costs questions. In light of the few but troubling warnings from industry and environmental activists that promotion of fast fashion creates incentives to cause serious ecological damage, the competition to this day remains on price.

―――――――

EXISTING IN ISOLATION IN THE PRODUCTION CHAIN?

Manufacturing of apparel is so fragmented that neither manufacturers nor customers understand much about how and when in the apparel production process environmental degradation occurs (Rosenthal, 2007). Therefore, ecological damage can occur anywhere in the making process, from the growing and harvesting of cotton, to the manufacturing of synthetic fibers, to washing, caring for, and disposing of garments. But by and large, ecological impacts are not addressed in the body of research (Morgan and Birtwistle, 2009). Among the reasons could be the fact that the overwhelming majority of industry research is survey based with respect to customer satisfaction (Naderi, 2013).

However, it is unclear whether this is really the case, or just a convenient summation of the apparent complexity of the industry. To outsiders, and even to production management specialists, the separate production links in the apparel value chain may appear to exist in relative isolation from each other. Perhaps for these reasons, there is a notable lack of integrative research into the whole system. The research foci are clearly divided into main categories in:

1. Input analysis: the literature on textile production, mostly covered by agricultural economics, agronomy, trade regulation, and technology advancement disciplines that focus on the growing and trade of cotton and weaving of fibers (Anderson and Valenzuela, 2007; Baffes, 2005, 2007, 2009; Delpeuch, 2007; Minot and Daniels, 2005). Most of the works in this field rarely mention demand as anything but external to the particular work, much less offer insight into factors that drive demand. On a few occasions a concluding note is made that demand has been impacted by globalization, but it is seldom described how.

2. Fashion production and operations management: the literature on fashion buying and merchandising, brand proliferation, brand extension, logistics, and integrated marketing (Cachon and Swinney, 2011; Park et al., 2012; Park and Rabolt, 2009; Pentecost and Andrews, 2010; Phao and Lo, 2004; Wang and Wang, 2008; Whitelock and Fastoso, 2007; Willems et al., 2012). A notable lack of textile production mentioned in these bodies of research indicates that apart from praising low prices, understanding of input-specific operations is not needed. All these bodies of research indicate in their assumptions that the main knowledge operations management needs to possess about fabric concerns fabric availability in its finished state. Fabric needs to be inexpensive and to come in many varieties. How it is manufactured is not the concern of production managers.

3. Fashion design: the literature on fashion categories, fashion innovation, fashion intellectual property rights, fashion design, and fashion promotion (Abecassis-Moedas, 2006; Naderi, 2013; Nenni et al., 2013; Pesendorfer, 1995; Tungate, 2012). The artistic body of fashion system research (although not exclusively; for example, Pesendorfer (1995), is a microeconomic paper), these fields seldom mention fabric production or retail merchandising. However, all offer insight into customer tastes and preferences.

The most extensive and varied category is Category 2, which falls into a larger academic discipline—that of operations and production management. Within this discipline, a surprising scarcity of fashion operations management work is evident. For example, the *Journal of Operations Management* (JOM) has only ever published 133 articles that even mention "fast fashion"—they include works on retail inventory management, general logistics, shipping, receiving, contracting, and marketing and customer relations. Particular topics are varied, as is the way that fashion as a phenomenon is examined. Meanwhile, it has published 207 articles with the word "automobile" in the title alone. *Production and Operations Management Science* (POMS) has published fewer than 125 articles on any aspect of fashion/garment/retail management. However, it has published close to 500 articles that examine "manufacturing" firms, grouping all industries that can be classified as manufacture intensive under the same prescription platform for what constitutes successful operations management strategy.

Among the over 400 works examined while writing this book, two authors offer an integrated look at the entire industry—Rivoli (2009) and Birnbaum (2005, 2008). In three books on the subject, the demand dynamic is examined in the context of promoting demand at the retail level and evaluating the impact of that promotion backward into the production chain. Three books out of 400 sources is not a good statistic. It is an example of why Rosenthal (2007) argues that the industry is too complex for most professional academics to study. However, academics are only one type of subject-matter experts. If academics stay vested in their silos of specialization and do not integrate the complexities of the entire system of fashion economics into their work, it does not mean that professionals within the industry do not understand its intricacies. They do, and there is a body of literature that tracks the nature of that understanding. It is the literature on category management.

Category management with respect to fast fashion explains that successful retailing requires integrated communication flow within the entire production system (Dewsnap and Hart, 2004; MacCarthy and Jayarathne, 2010; Sheridan et al., 2006; Walters, 2006; Willems et al., 2012). The conclusions stress that all major decision makers in the production chain are very aware of each other's operations. The higher the knowledge, the lower the transaction cost of communicating information in terms of volume.

Reviewing the legacies of research in fashion supply chain management, Wigley and Provelengiou (2011, p. 142) show the "management of relationships" both in "back-of-house," meaning the production end, and "market-facing," meaning the retail end of the supply chain. This is the case because of the need for entire system competency, which can only be built through fostering managerial relationships that extend up the supply chain to designer, supplier, and manufacturer and down to distributor, retailer, and consumer. Therefore, as Sheridan et al. (2006) explain, marketing and advertising decision makers know every aspect of the production and transportation operations of their preferred garment manufacturers. These aspects include awareness of fabric and trim sourcing partnerships as well as transportation and clearance intermediaries. Therefore, marketing professionals know exactly how far each item travels and how long it stays at border crossings. They also know warehousing and holding capabilities because of the need to manage inventory backlogs.

Knowing which items move relatively more slowly and create costly inventory piles is essential in promoting their alternative movement. So, to say that industry operatives are not aware of the ecological impact of their actions is naïve. They are aware, but they treat them as direct costs. Most importantly, they communicate them as such to the end consumer. The message is that through superior operational knowledge, they can lower the direct costs of production and in that way, pass the lower costs down to the consumer. This is the spiral that spins fashion downward from high end to fast fashion; a spiral of managerially lowering direct costs and promoting the success of that process to the consumer via the introduction and celebration of low prices. It is in this way that the industry hides the uncomfortable truth that growing demand carries secrets of high pollution.

The academic community does not question this silence on fashion consumption's ecological impacts, because unlike industry professionals, academics truly exist in isolation in the scholarly chain of fashion merchandising and manufacturing research. Therefore, the ecological damage that occurs at every stage in the garment-making process—from the growing and harvesting of cotton, to the manufacturing of natural and synthetic fibers, to the manufacturing of finished clothes, to their distribution and sale—is not discussed.

What has garnered attention has been the treatment of clothes post purchase, primarily their washing, care, and disposal. Cline (2012) focuses on that end stage of garment use, while noting that prior stages also contribute to environmental issues. The examples offered are mostly on the use of water, from agricultural water use to washing and caring for the garments. Levi's jeans also rallied around the water issue in introducing their line of sustainable products; in their case, "sustainable" meant encouraging lower water use by asking their customers not to launder jeans, that is, the "no wash jeans." The problem is that the responsibility in this end stage is put on the end consumer and removed from the production players. As already discussed, it is a retailer-driven industry. As long as the incentives in the production chain are free of ecological scrutiny, the understanding of production efficiency will not include environmental impacts. And the convenient excuse will continue to be that the system is too complex to understand.

6

The Direct and Social Costs of Low Prices

THE SOCIAL COST OF FIRM PROFITABILITY

People seldom remember that once clothes are sold, their life span does not come to an end. Rather, a whole new economic chain of used clothes commerce begins. Its volumes have been noted in policy and ecological discussion because they impact entire economic development trajectories in poor countries.

As noted, Cline (2012) explains that according to the Environmental Protection Agency (EPA), Americans throw away about 70 pounds of textile product a year. Further, Americans donate such an overwhelming amount of clothes to charity that more than half are not needed, but are instead sold directly to textile recyclers. These recyclers do not recycle the clothes, but in turn sell them to overseas entrepreneurs for repurposing as secondhand clothing. Much of this "recycled" clothing ends up in Africa, not as donations, but as direct sales. This dynamic plays out not only in America but also in the West in general.

The export of used clothing to developing nations has generated a whole body of research because of the impact of those imports on competitiveness issues. Since the early 1970s, Western secondhand clothes exports to developing nations have been in such volumes as to threaten the development of local apparel sectors. In response, some developing nation governments have banned used-clothing imports in order to offer protection to their domestic textile industries (Haggblade, 1990). Hansen (1999) explains that the amounts of these exports are such that by the mid-1990s, used clothes comprised the sixth largest export of the United States to the entire region of sub-Saharan Africa. This statistic is staggering because it refers to all exports, including commodities, foodstuffs, machinery, commercial equipment, and, as was discussed in Chapter 5, raw cotton.

Studying the impact of secondhand apparel commercialization in the era of fast fashion, Brooks and Simon (2012) show that there are different market structures in used clothes commerce. In addition to formal markets where sales are executed in consignment stores, most of the commercial activity happens in informal markets, also referred to as black or gray markets. These are markets that, for various reasons, are not regulated by legal institutions. In developing nations, such as the sub-African nations mentioned in the literature on recycled apparel, informal market structures exist in most commercial activity (Claudio, 2007; Hansen, 1999, 2004). With respect to the trade of secondhand clothing, informal markets define not only the sales of the individual items but also their transportation and distribution, because although many countries have banned the importation of used clothes, they are still smuggled illegally to fill gaps in local apparel consumption that the domestic apparel industries are not addressing. Brooks and Simon (2012) show how in the apparel sector there is a whole and very powerful illegal import trade in poor African nations. The legality issues importers of used clothes are skirting stem from, on the one hand, the preferential trade agreements discussed earlier that support African apparel exports, but, on the other, the failure to address gaps in local apparel markets in terms of consumption. As the apparel sectors in Africa, in both textile and garment manufacturing, grew, they grew with respect to exports. However, the local (be they foreign-owned, domestic, or joint venture) producers did not focus on producing clothes for the local consumer. Therefore, incentives kept increasing to import used clothing for local consumption.

These incentives are high to begin with because used clothing has no production costs. Its commerce is based on minimal repurposing and, therefore, the markups in profit making are significant. Abimbola (2012) describes the process of repurposing, which is done by Western suppliers.* The clothes are sorted in a sorting factory, which is a facility that pretreats donated or discarded apparel, classifies it, and sells it to merchants. The sale is promoted by the commercial owners of repurposing facilities based on the quality of the garments. The next step is for an independent importer to buy the bales and in turn resell them to individual shops in developing nations; in effect, continuing the story of the geographic peculiarities of

---

* In the case of that article, the suppliers are British.

used clothing commerce, Walther (2014) explains that most of it is done in Africa. There, a whole evolved and highly complicated border-crossings process ensues, because importers face two main challenges to reaching the desired markets for used clothing. One is the lack of physical infrastructure for storage and transportation, but another is the presence of market-protecting policies in certain nations that ban the importation of used clothes. The result is a complicated and long transportation chain that creates its own subcommerce in the distribution of used clothing.

The above-explained dynamic is important to understand because it illustrates why the commerce of used clothes is important in the local economic dynamics of developing nations. It supports a retail-like economic activity that addresses a market gap in apparel (the same scenario is also the case in the sale of used shoes) for the average citizen. In developing nations, where commerce is based on average income of few dollars a day, discretionary income brackets only allow purchasing items that retail for the equivalent of American pennies. Discretionary spending can be defined as the percentage of income the consumer has left for goods and services after taxes are paid and loan obligations are met. By definition, because of poverty and institutional underdevelopment, people in very poor countries do not have tax and loan burdens to service. Their entire incomes go toward sustenance and shelter expenses, including clothing.

Because most people in developing nations earn pennies a day, the amount of money they can spend on necessities such as food and clothing is minimal. The nature of such restrictions on spending options in relation to clothing purchases is thoroughly discussed in the context of secondhand clothing commerce by Baden and Barber (2005). The authors offer the example of Ghana, where they posit that because of general poverty levels, 90% of the population rely on secondhand clothing. The clothing sold there, and in Africa in general, is not in retail facilities but in street markets, on street corners, and at trading posts (Hansen, 2000; Brooks, 2012). It is much needed and sporadically available due to the dynamics of shipping and transportation costs discussed by Walter (2014) and Abimbola (2012). Hanson (2000) explains that in Zambia, there is a term to describe the local used-clothing buying experience. It is *saluala*, which means rummaging through piles in a frantic manner.

Due to these facts, in very underdeveloped nations it is used clothing, shoes, and outerwear that meet the apparel consumption needs of the average person (Abimbola, 2012; Brooks, 2012, 2013; Brooks and Simon,

2012; Hanson, 1999, 2000, 2004; Wydick et al., 2014). Therefore, it is very unprofitable for local clothing manufacturers to engage in any new clothes production for local mass-market consumption. The demand for used clothing in very underdeveloped nations is so high mainly for these reasons. It is also high because of the fact that the used clothing is of Western brands (Baden and Barber, 2005). This is a cultural feature of demand that is related to the desire to belong to a global culture. As explained by the literature examined in Chapters 1 and 2 on global brand proliferation, people in the developing world exhibit a preference for globally recognized brands, because possession of these signals social success higher than their natural and relatively poor environments.

The fears of market distortion from unfair competitive pressures on local producers might be unfounded, because both markets remain decidedly separate, particularly in relatively poorer nations. The poorer the nation, the lower the purchasing power of its average consumer, and often that purchasing power is below the formal market threshold to bear even the most minimal production costs in local manufacturing. Therefore, it is not profitable to sell originally manufactured goods in a wide-scale local-consumption format. Wydick et al. (2014) offer evidence of this dynamic in rural El Salvador. The authors investigate whether shoe donations negatively impact the commerce of local shoemakers, and find that they do not, because local shoe manufacturers rely mainly on exporting their products. Therefore, local shoe-market gaps remain, and are addressed in the donation-driven secondhand informal market.

The commerce of used apparel continues to grow as Western shoppers increase the frequency of clothing and shoes purchases. As Cline (2012) carefully tracks, and much of this book's data on sales volumes of fast-fashion conglomerates indicates, the amount of clothing the average Western shopper purchases in one year is several times what it was in the 1990s. The amount of clothing the average Western shopper discards is proportional. The result is a drastic increase in the volume of trade in used clothing (Norris, 2010).*

---

* While making this point, Norris (2010) puts it in the context of India. She argues that the increase in imported used apparel in India is due to the advent of fast fashion in Europe and its notable embrace by Britain. The historic close trade relationship between Britain and India leads to the high concentration of British exports of used apparel to India in relation to other exporting nations.

The United States is the largest exporter, and dispatches over 500,000 tons of used clothes annually to more than 100 different countries (Rivoli, 2009, pp. 216–17). The second largest is the United Kingdom, which, according to Brooks (2013, p. 2), exported 319,998 tons in 2011. According to Norris (2010), most of the exported British clothing ends up in India. According to Hansen (1999), most of the exported American clothing ends up in Africa. These are very important statistics, because they illustrate the direct link between fast fashion and the growth of fashion waste.

The American averages are related to a market dominated by traditional retail, with a relatively lower concentration of fast-fashion conglomerates in relation to branded retailers. For example, Zara has just recently entered the American market, and only in large metropolitan areas. Fashion experts posit that its corporate parent Inditex is not interested in the significant proliferation of Zara stores within the United States because H&M has already established itself as the leading fast-fashion chain.* In addition, American retailers such as Gap and J.C. Penney are aggressively developing their own fast-fashion strategies. For example, J.C. Penney announced its "no more sale" strategy on January 25, 2012 (D'Innocenzio, 2012). These are fairly new developments in the traditionally branded retailer-dominated American market, and it is unclear from the data on used apparel exports whether the frequency of discarded clothes in America has increased significantly in relation to its historic high volumes. After all, the arguments by Haggblade (1990) and Hansen (1999) that America is the largest exporter of used clothing come from the time period before the advent and proliferation of fast fashion.

Another fact that must be noted in relation to this evaluation is that in America, 316 million people are responsible for the discarding of apparel that can be repurposed and resold to the tune of 500,000 tons. In Britain, only 64 million people are behind the exportation of 320,000 tons of used clothes. This fact is direct evidence of how much fast fashion has increased the incidence of discarding apparel, because the British market today is dominated by fast-fashion retailers, not traditional branded retailers (Barn and Lea-Greenway, 2006; Bruce and Daly, 2006; Bruce et al., 2004).

Sales in Britain are so strong that Rosenthal (2007) posits it is the birthplace of fast-fashion success. The author offers the statistics that between 2001 and 2005, the sales of women's clothing alone there rose by 21%,

---

* For a discussion on Zara's strategy in market assessment, refer to Lopez and Fan (2009).

as the British market embraced the fast-fashion model more than any other European market, including those of Spain and Sweden—the home nations of the fast-fashion leaders Zara and H&M.

The amount of clothes the average Western shopper discards through either direct throwaway or donation is subject to sustainability scrutiny from social activists such as Cline (2012), who explains that it was the carbon footprint of apparel waste that changed her from a self-proclaimed *fashionista* to an environmental advocate. There are many reasons, as already explored in this book, for the high ecological damage of the production of new clothes. But it was the fashion waste that sparked the scrutiny of the ecologically minded analysts.

There are two main reasons why apparel waste, rather than the production of new clothes, is the focus of inquiry. One is the long supply chain of a garment life cycle. From the harvesting of cotton, to the manufacturing of fabric, to the assembly of clothes, to their sale, to their repurposing and resale, transporting apparel around the world creates significant environmental damage because it is mostly done via commercial shipping, which is among the highest carbon-emission-intensive modes of transportation. But the second, and arguably more important, reason, though less discussed in the literature, is that the final disposal of used clothes disproportionately occurs in the developing world, where there are few, if any, regulatory standards in waste management (Ahmed and Ali, 2004; Henry et al., 2006; Thomas-Hope, 1998). The result is that, in terms of toxicity, the two most ecologically damaging phases of the clothing life cycle disproportionately happen in the developing world. The first phase is the production of the clothes, which, as explained in Chapter 5, is the most toxic and least addressed ecological problem in fashion economics. The second phase is the final discarding of used clothes, which decompose in developing world soil and waterways (Vergara and Tchobanoglous, 2012).

Because of these reasons, activists such as Claudio (2004), Cline (2012), Rosenthal (2007), Brooks (2012, 2013), and Hanson (1999, 2000, 2004) have called attention to the ethical aspects of promoting the overconsumption of cheap clothes. It is the overconsumption, meaning consuming without clear need to satisfy, that leads to the high incidence of discarding apparel, which creates the explained notable amounts of fashion waste. The calls to reexamine cultural aspects of overconsumption are echoed in other industries as well. Rising awareness of global environmental and social problems, mainly accelerated by rising consumer protection group pressures, has forced global companies* to recognize these demands in their production processes and investment strategies (Gray, 2006).†

As MNCs have started to become more powerful worldwide, it has been generally argued that they have a social responsibility to operate ethically. Corporate social responsibility (CSR) has become a global slogan (Carroll, 1999; Lee, 2008). Activists have called for more stringent rules and regulations to be issued and enforced (Matten and Moon, 2008; Porter and Kramer, 2006). Concepts such as sustainability accounting and reporting (SEA) have been introduced. SEA is an accounting measure for shareholder value that includes sustainable development components (Gray, 2006). In fashion and other fast-moving consumer good fields, large international buyers have also implemented their own codes of corporate ethics. Noncompliant suppliers are pressured to reconsider their own standards and streamline their operation policies with the corporate ethics policies of their clients (Diebäcker, 2000). A multitude of such corporate ethics codes have emerged, including model codes drafted by trade unions and nongovernmental organizations (NGOs), company codes, and government-promoted codes.

In the apparel industry, codes of conduct are the subject of debate, for they have different meanings for different people. For some, they are a way of avoiding binding regulations. For others, they are a means of addressing a regulatory gap, which is often temporary (Nimon and Beghin, 1999). They are also a way to stave off more demanding regulation by encouraging soft laws. At the same time, many of these codes operate in isolation, which can create confusion.‡ In apparel, Europe is the leader in implementing programs such as Germany's Eco-Tex Standard and the European Largest Textile and Apparel Companies (ELTAC) eco-label initiative (Birnbaum, 2005).

As early as 1993, Germany introduced two eco-labels: (1) *Markenzeichen Schastoffgeprufth Textilien* (MST), which set norms for consumer goods and indicated a lower content of pollutants, and (2) *Markenzeichen*

---

* It should be underlined that most MNCs that have been under attack are headquartered in the West. The literature as a whole addresses "Western" companies in its charges of environmental exploitation. However, as seen in Chapter 5, many of today's leading apparel MNCs are developing-nation firms. It is unclear whether they face the same level of scrutiny as Western MNCs.

† The cited work by Gray (2006) refers specifically to apparel MNCs, the majority of which were based in the West. However, as explained in Chapter 4, many of today's leading fashion and other fast-moving consumer good MNCs are based in the East.

‡ For more information on codes of conduct initiatives, refer to the Clean Clothes Campaign reference guide at cleanclothes.org.

Unweltschonende Textilien (MUT), which sets norms for production processes. MUT indicates that all processing conditions were analyzed with reference to the degree of pollution of air, water, and soil. More recently, Germany has introduced the Eco-Tex Standard. It lists various criteria for evaluating textiles from an ecological perspective. The Eco-Tex standard has been developed by the International Association for Research in Eco-Textiles, and offers manufacturers of garments and textiles an opportunity for certification of eco-friendly products. As a result of these regulatory policies, the use of certain dyestuffs, such as cobalt blue and sulfur black, has been totally banned in Germany (Birmbaum, 2005). Wider Europe is following in the footsteps of Germany, but there is resistance from certain nations because of fears that retail sales may suffer if choice and variety of product options are limited. Therefore, in wider Europe, textiles dyed with cobalt blue and sulfur black are not banned. Each country has discretion on how to implement eco-standards. Richer nations are more stringent than nations such as Spain and Italy, where most of Europe's extent apparel manufacturing is concentrated.

Toxicity in textile production can hardly be addressed by the banning of two dyes. Therefore, codes such as MST and MUT only provide a false sense of security for fashion buyers. Furthermore, when it comes to the fashion industry, and the Eco-Tex and MUT certificate programs in particular, the overwhelming majority of standards outlined in their codes of conduct have to do with the ethical treatment of workers, which is an important issue.

With respect to working conditions, Pan and Holland (2006) note that most textile manufacturing facilities are classified as "small and medium" size because they employ relatively few workers—40–200—and export to more than 50 branded retailers. Those branded retailers outsource the sewing and finishing of garments to garment manufacturers that are considered "large" in industrial size because they employ from 700 to as many as 3000 people. Working conditions in both facilities are different. In a textile plant, the health hazards workers face are usually noise and noise pollution. In a sewing factory, all the horrors that are described in the recent Bangladesh factory fire are present. So, when meeting voluntary codes of conduct, a textile facility could implement fairly few working condition improvements and show relatively good working conditions in terms of air quality, facilities, and even workload. This would earn it high marks. However, that same factory could continue unchecked to emit toxic effluents if no regulatory structures are enforced. Therefore, a voluntary code of compliance with labor standards does not include environmental regulatory compliance.

Summarizing the extant research on ethical consumption, Joergens (2006) states that there is much interest in ethical fashion as a concept, but it is unclear whether customers would sacrifice their own bottom line for a larger collective good. As Beard (2008) explains, there has been interest in eco-fashion, or ethical fashion, because eco-fashion increases the utility level of fashion purchases. Therefore, eco-fashion has social impact implications. Customers feel better about the purchase if they believe it influences more than just their own wellbeing. To that effect, making eco-fashion purchases has sociopolitical aspects (Gam, 2011). Because environmental stewardship falls within political platforms along party lines, buying eco-fashion has become a political statement (D'Souza, 2015).

Despite all this, eco-fashion sales are still the anomaly. With all the interest the concepts of eco-fashion and sustainable consumption are generating among academics and social activists, business evidence abounds that the temptation of low prices, paired with the societal pressures to be stylish, dominates the purchasing behavior of even the most environmentally conscious fashion buyers.

UNDERSTANDING CUSTOMER UTILITY

Gabrielli et al. (2013) argue that industry insiders have questioned the viability and sustainability of the fast-fashion phenomenon, but that it is time to accept the reality that fast fashion is not only here to stay but has become the pillar of fashion consumption for the average buyer of clothes. The authors find that elements of fast-fashion image, consumption, and behavior are present as part of individuals' daily lives. The proliferation is based on the high value that customers get from the increased choice of clothing options fast fashion provides, the increased utility level of feeling fashionable when changing clothes often, and the low prices that allow shoppers to afford varied and repeat purchases. Utility in particular, the concept of gaining satisfaction from a purchase, has been of much interest to fast-fashion retail scholars.

Choi et al. (2010) examine factors that influence consumer preferences toward fast-fashion brands, and find that perceived quality and product feature similarity ranked high in terms of evaluating utility. The key point is in the language of analysis, referring to "perceived" quality. As Laura Heller of *Forbes* magazine explains (Heller, 2014), when dresses are $8, camisoles $1.80, jeans start at $7.50, and the demand for clothes priced below $2.00 an item is growing, what role does quality really play? Despite this fact, the traditional consumer-choice fashion research examined in Chapters 1 and 2 has evolved around the concept of quality. However, more and more current research puts the quality of the garment fairly low on brand factor judgment scales.

Fiore (2010) succinctly states that in fashion, customers no longer depend on quality to differentiate products. This is not a new finding; Fiore cites Pine and Gilmore (1999), who had made the same conclusion a decade earlier based on new information on production improvements that allowed equalization of quality, not only in fashion but across sectors. The explanation is that machines were getting so good as to be virtually unable to produce bad-quality inputs. In that context, bad quality would be related to defects and management's decision to purposefully sell defective products. However, even with exceptions, the incentives to engage in such behavior are low, because consumers face many substitution choices, and bad customer experiences that result from the purchase of defective products send shoppers to competitors.

Undoubtedly, this demotion of quality as a fashion attribute denotes that fast-fashion quality has become increasingly subjective, because the traditional definitions of durability that have defined the understanding of high quality cannot be applied to readily disposable garments that are designed to be discarded before they wear out. For example, Kim (2012) studies how individual customers rate the importance of garment quality in the context of their overall brand experience. The author argues that for customers, brand experience is dimensional, and offers a conceptual framework to evaluate it. Quality attributes, as traditional fashion categorization would dictate, are performance, reliability, style, and design. The author puts these attributes in the first stage of the framework, entitled "brand identify/meaning," which is the lowest stage. The higher stages are credited with more importance in satisfying vital experiential needs. In the second stage, entitled "brand response," the experience is separated into "judgment" and "feeling," and it is in "judgment" that "perceived quality" is placed. But, as per Kim's (2012) recommendation and logic, Stage 1 level of satisfactory experience must be completed before moving on to Stage 2. The question arises, then: why is "performance" separate from "quality?" Is the implication that consumers rate performance differently from quality? And how can they be separated in a garment? Can a high-quality garment not offer high performance? And what exactly is the meaning of performance of a piece of clothing? Perhaps due to the lack of clarity on such questions, the discipline research is disjointed and, as Beard (2008) states, confusing. The lack of clarity is contingent on the fact that the concepts are seemingly self-explanatory. However, they can be ambiguous, with the result that they can mean different things to different people, and can also be weighed differently in the individual decision-making processes shoppers make.

To address such difficulties of consistency and clarity in categorical classifications of the customer behavior of fashion buyers, one can turn to the literature on brand association (Bridges et al., 2000; Janiszewski and Van Osselaer, 2000). Although brand association research is general in terms of industrial focus, much of it is based on fast-moving consumer goods, including clothes, accessories, and cosmetics. This body of research looks at all brands of consumer goods and offers insight into integrated marketing strategies for developing brand awareness. Although the main conclusions of the brand association field of research are not put in the context of fashion, understanding brand association can be useful for understanding utility of fashion purchases because of the implications for brand extensions. Brand extensions are launches of new product lines under the umbrella of a well-established brand. Having similar features to the concept of product diversification, brand extensions differ in the fact that they rely on higher levels of product similarity and complementarity.

Offering opportunities for diversification mainly along price lines, brand extensions allow product positioning at different price points. The core link between the brand extension's successful operations management and brand association is that utility of purchase rests with association. Brand association relates to anything that describes brand likability (James, 2005). Put simply, customers are predisposed to know how a purchase will make them feel and how it will result in a change in utility levels prior to the actual purchase. This is based on past experiences and building an attitude about expected satisfaction. The stronger the brand association, the more the latitude of feeling shrinks. Therefore, the more a customer likes a brand, the more the marginal utility—that is, additional, new, and unexpected level of satisfaction—shrinks. This means that the utility functions of branded fashion purchasers are fairly inelastic.

An inelastic utility function denotes lower propensity to switch brands, or, as it is described in fashion terminology, to change brand allegiances.

Customers become self-identified with brands, and this is why Kim (2012) offers Stage 3 of the brand experience, in which brand relationship is measured through constructs that include behavioral loyalty and attitudinal attachment. These two concepts stand out in the vast body of fashion merchandise research, because they denote the nature of growth in fast-fashion sales, which is to institute a mentality of repeat purchasing.

Since in fast fashion, prices are kept low, the possibility of extracting fashion premiums is absent. Kim (2012) also includes premium price in Stage 3 as an outcome of a Stage 3 brand relationship, but in that model the reference is to fashion overall, including high-fashion brands—those that most rely on markups and are positioned at prestige-market price levels. In fast fashion, behavioral loyalty would be manifested as habitual shopping. Habitual shoppers have inelastic consumption functions, and it is this fact that has important implications for theoretical economic analysis of fashion sales. The implications affect utility measures because elasticity of consumption indicates how customers perceive the value of a purchase.

Customer utility is defined by value derived from the amount of money spent on an item in relation to discretionary spending. In the case of clothes, the average American spends a negligible amount of total expenditures on apparel. According to the American Bureau of Labor Statistics (BEA),* which offers expenditure data per household measured in consumer units, the average American household consumes $51,000 worth of goods and services in a calendar year as measured by household discretionary spending in relation to household disposable income. On average, for the past few years,† the BEA calculates the amount allotted to clothing purchases to be around $1,600. In comparison, Americans spend about $8,000 on food, $18,000 on shelter, and upward of $5,000 on health care. These are the largest, as it is put in economic jargon, discretionary spending brackets. As previously noted, discretionary spending is the amount of money people have left to spend as they choose after taxes, legal liabilities (such as debt obligations), and "necessities" are paid—food, shelter, and clothing. The problem is that, as this book shows, today in America, and the general West, clothing purchases are not made out of necessity. This fact calls for reconsideration of the theoretical assumptions of consumer theory in the context of fashion economics.

---

* Table A shows these averages, available at: http://www.bls.gov/news.release/cesan.nr0.htm
† Table A provides values for 2011, 2012, and 2013. Data from earlier years are available through a targeted query. The values remain remarkably similar.

THEORETICAL IMPLICATIONS FOR ECONOMIC ANALYSIS: ELASTIC CHOICE BUT INELASTIC SALES

Consumption theory states that inelastic utility functions lower the probability of substitution, particularly when the question of price is settled (Hicks, 1970). Based on that definition and the examined utility factors associated with fashion purchases today, one can conclude that sales in the industry should tend to be inelastic. This outcome, by definition, is the whole point of advertising—to make the consumption function of buyers more inelastic over time. Where, historically, such decrease in elasticity allowed producers the ability to charge higher premiums, also explained through the concept of extracting economic rents, today, with respect to fast fashion, prices are lowered below rent-extracting levels.* In this context, inelastic consumption leads to lower propensity for product substitution, but higher propensity to increase the volume of product purchases in response to increased levels of innovation. However, traditional fashion economic theory rests on the assumption that apparel sales are elastic.

Most famously, Krugman and Helpman (1985) model utility of apparel consumption and posit that apparel purchases are, at their core, elastic. The authors assume that the elasticity is due to a high variety of choice and the fact that this variety, subject to budget constraints, spreads across multiple individual goods. In other words, apparel shoppers do not buy items in an isolated decision context. They purchase clothing items that can be worn with other items, therefore exhibiting high propensity for change in choice. It is these assumptions of choice and the ability to change that have shaped the theoretical work on apparel sales as one defined by high elasticity of both consumption and price. However, the entire point of the customer service-oriented management bodies of literature within the few marketing disciplines dedicated to advertising and promotion is to explain how producers, in effect, combat choice.

Choice is a threat. The more choices customers face, the higher the competitive pressure producers face, and the higher becomes customer propensity to lower consumption of particular brands. However, there is no evidence of such a dynamic in fashion sales. For that reason, contrary to traditional economic-theoretical assumptions, when it comes to the aggregate industrial reality of apparel consumption, the elasticities are

---

* Economic rents are to be explained shortly in the context of price markups for fashion goods.

different because they are not subject to decrease in the aggregate number of purchases.

Today, on average, fashion shoppers may face many choices when making a purchase, but their overall buying habits exhibit an inelastic and gradually increasing consumption function. It is inelastic because there is no evidence that economic pressures significantly affect the annual apparel expenditures of the average person. It is increasing because of the amount of clothing the average person purchases on an annual basis. As already noted, this amount has grown exponentially in the recent past, unencumbered by economic fluctuations such as the Great Recession of 2008. The reason is that, partly due to fast fashion's advent, apparel prices have been lowered to such unprecedented retail levels that they are arguably not subject to significant budget constraints.

Bijmolt et al. (2005) point that pricing is one of the most important issues in marketing and offer a meta-analysis of price elasticities for 1851 branded products tracked in 81 academic studies. Among these products are quite a few fashion and related goods. The authors posit that technological advances and societal changes have altered the economic realities of market characteristics of product categories with respect to branding. The argument is that prices are becoming increasingly elastic because of growing competition in all industries and general improvements in productivity. The authors propose that the following factors define price elasticity: (1) time trend; (2) manufacture brand versus private label; (3) product category; (4) stage of product life cycle; (5) country (both of origin of product and retail environment); (6) household disposable income; and (7) inflation rate (Bijmolt et al. 2005, p. 142).

All these characteristics, as discussed in previous chapters, have changed in fashion economics. Time trends are undefined, and it seems from the constantly downward-spiraling pricing competition in fast-fashion retailers that trends do not offer higher-price markup options. Manufacture brands versus private label is the core change in pricing that has allowed fast-fashion retailers to compete with the designer brands of the past. Furthermore, the branded retailers of the 1990s and today challenge the private-label pricing model because they offer their own "manufacture" brands, but they also promote their image in equal part with the merits of "private labels," as Bijmolt et al. (2005) put it, which they also sell in their stores. The third pricing factor listed—product category—has become blurred in apparel. As Chapter 1 explains, the historic delineation of fashion goods on a category scale, from mass market to luxury, has been blurred and is converging. This fact is very important for understanding the price elasticity of fashion goods, because the assumption is that during the competitive process, producers can adjust price based on category demand. If the demand for luxury items decreases, then their price will decrease to the point of reclassification. However, because of branding, this dynamic seldom plays out in modern apparel retail.

The fourth factor—stage of product life cycle—is just as nebulous, because it is based on trend promotion and the changeability of the public's perception of what is stylish, what is fashionable, and what is both stylish and fashionable. The last three factors are all economic conditions that would denote price elasticities at the production level—country of origin, household disposable income, and inflation rate. In general, these production-level elasticities can lead to elasticity of consumer spending; however, in fashion that is not the case. As seen by the cumulative summary of modern fashion retail offered so far in this book, the impact of these production factors that are supposed to increase price elasticity is skillfully mitigated by fashion manufacturers through the one thing that they are supposed to define—price. By lowering prices to the point that they fall into the lowest disposable income brackets, fashion retailers defy elasticity pressures that stem from economic constraints. Therefore, in fashion sales:

1. Inflation is not a major elasticity concern. The assumption is that it would be if its fluctuation influenced production costs and, therefore, could lead to a rise in prices, but as explained by Chapters 3, 4, and 5, production is so highly subsidized that inflation can easily be offset by the price support system in international textile and garment production. As explained, textile manufacturing is the core production component in the final value (and price) of a garment. It is also the most subsidized. The level of subsidy is so high, as examined in Chapters 3, 4, and 5, that it would be unwise to think that modern-day levels of possible inflation fluctuation can impact the prices of the global textile market. In addition, one must remember that textile production is concentrated in nations with high inflation problems. The fact that it has grown in those nations in the past two decades, despite market uncertainties, political and social instability, and general economic problems, indicates that inflation is not a defining contributor to price and consequently, sale elasticity of apparel products.

   Inflation can also impact other direct costs in the manufacturing process, such as transportation costs and manufacturing costs. As explained in Chapter 1, transportation is the second most costly factor in the garment supply chain. The evidence provided in Chapter 3 on shifting supply chains indicates that apparel producers efficiently mitigate transportation costs. The evidence is that even with the fairly recent price volatility of fuel and energy costs, fashion retail prices keep on decreasing. Therefore, higher transportation costs or inflationary pressures in energy markets do not seem to impact price elasticity. Additionally, as shown in Table 1.1, in fashion production (both traditional and fast fashion), manufacturing costs, in terms of assembly, comprise a mere 6% of total production costs. This fact, paired with the reality that most direct manufacturing is completed in nations with relatively high inflation and in the context that 6% is a negligible part of total costs, indicates that inflation at the manufacturing stage is also not a contributor to price elasticity.

2. Country of origin, Bijmolt et al. (2005) posit, is supposed to affect price elasticity with respect to customer preference. It is assumed that certain products from certain nations would be more valuable (therefore, their price less elastic), such as electronics from Japan and automobiles from Germany. However, in fashion that is not the case, and it is unclear if it ever was, because of the quota system discussed in Chapters 3 and 4. It is true that for certain couture and other high-end and luxury items, in the past the label "made in Italy" or "made in France" mattered. But today, even in high-end fashion product lines, the traditional "made in Italy" has been replaced by "made in China," and buyers do not care (Bertoli and Resciniti, 2012; Tokatli, 2014; also see Donadio, 2010). Tokatli (2014) explains that Prada has changed its economic-geography strategy to manufacture mostly in China, its labels say so, and its sales are growing despite this fact.

As noted in Chapters 3 and 4, neither the general economic malaise of the developing world, where most apparel production has shifted, nor the economic stagnation of the developed world market has impacted the unprecedented growth of apparel sales. Therefore, "country of origin" does not matter at all in fashion economics. Although Gereffi and Frederick (2010, p. 2) state that the 2008 global financial crises and subsequent recession "hit the apparel industry especially hard, leading to factory shutdowns, sharp increases in unemployment, and growing concerns over social unrest as displaced workers sought new jobs," there is little evidence that the aggregate industrial sales truly suffered. As explained in the data analyzed in Chapter 1 on profit and retail-outlet growth post 2008, for fast-fashion conglomerates, which dominate the overall industrial makeup of the apparel sector, sales have continued to grow at unprecedented rates.

3. Household disposable income matters in fast fashion, but for precisely the opposite reasons to those listed by Bijmolt et al. (2005). Their claim is that high disposable income should be associated with low sensitivity to price, and therefore, low elasticity. Low disposable income should be associated with high sensitivity to price, and therefore, high elasticity. However, from the outset, fast-fashion strategists specifically targeted apparel consumers with low disposable income.* The fact that fast fashion was born in countries with relatively lower discretionary spending brackets, and continued growing in terms of sales during the recent global recession, shows that its sales are inelastic when subjected to higher budget constraints. It is important to note once again the difference between disposable income—the income left after taxes—and discretionary income—the income left after paying taxes, interest on debt, and all essential needs. It is possible to have high disposable, but low discretionary, income. This is the case in West European nations. As explained in 'The Social Cost of Firm Profitability' at the beginning of this chapter, fast-fashion retail was particularly embraced in the United Kingdom, and Britain became the main market for fast-fashion conglomerates. Discretionary spending brackets there are some of the lowest in the developed world due to the high cost of living, high taxes, and the high prices of British retailers (Thompson and Thompson, 2009; Toynbee, 2003). Therefore, the average Brit does not have as much money as the average American (to offer the most often-used comparison) to allot to clothing purchases. For such reasons, the low-price strategy of fast-fashion promotion is attractive to UK consumers in particular (Bae and May-Plumlee, 2005; Bruce and Daly, 2006, 2011; Bruce et al., 2004; Byun and Sternquist, 2011; Doyle et al., 2006, 2005; Tokatli et al., 2008). Because of these facts, it is discretionary, not disposable, income that must be included in fashion economics models.

---

* Chapter 7 of this book focuses on this fact by examining the reliance of fast fashion on young consumers, who are by definition nonwage earners, which makes their disposable income levels very low in sociodemographic comparison.

Discretionary spending bracket fluctuations are important to relate to price elasticity because the price-positioning strategies of retailers depend on the size of discretionary spending brackets of different market segments. In other words, production managers position their goods at price points most closely aligned with their estimates of how much money the average customer can afford to spend on specific products. When economic conditions change, and their change is significant enough to impact discretionary spending, prices fluctuate. This fluctuation is a function of managerial decision-making. When discretionary spending decreases, retail managers must respond by lowering prices or positioning newly developed products at lower price points, as consumers cannot afford to spend the same amounts that they previously could. However, when discretionary spending brackets increase, production managers can increase prices, but they can also increase choice by offering new or complementary products at the same price points. It is this relationship between discretionary spending and choice that has implications for price elasticity, because traditional economic theory posits that increasing choice leads to decreasing prices because increasing choice leads to more competition. This is assumed to be the case because high numbers of choice options are supposed to lead to an increased customer propensity for substitution. This increased propensity for substitution, that is, willingness of consumers to try new choices, is supposed to lead to higher price elasticity, or price volatility. But in the case of clothing retail, these economic-theory assumptions need to be relaxed, because the market structures of the industry are defined by monopolistic competition (Krugman and Helpman, 1985).

Monopolistic markets are said to have lower levels of competition. Monopolistic prices are expected to have lower elasticity. Low price elasticity leads to low sales elasticity. This is the assumption because lower competition allows the owners of the monopoly to extract higher prices (or economic rents), and customers are expected to be willing to pay these relatively high prices because the low competition in the market results in few substitute choices.*

Another reason why relatively free markets can suffer from monopolistic-competition inefficiencies is related to transaction costs.† Monopolistic competition can arise when the transaction costs of information gathering

---

* These assumptions are empirically analyzed by the seminal work of Dixit and Stiglitz (1977).

† Neoclassical economic theory states that monopolistic market structures are inefficient because they lead to loss of both producer and consumer surplus. For a thorough discussion, see Perloff (2010), Chapter 9.

and processing in choosing possible product substitutions are high. In modern-day fashion, this is the case because of the high transaction costs of making the "right" fashion choice that arise due to increasing information-gathering and processing costs. Information-processing costs increase as a function of advertising. The more advertising inundation the customer faces, the more information distortion there is, and the higher the transaction costs of reconciling the distortion, processing the information, and making the optimal (in terms of utility) fashion purchase.

Perloff (2010) explains that in advertising-heavy industries with markets saturated by choice of substitute products, elasticity of consumption decreases as consumption becomes habitual. In these markets, advertising becomes increasingly important because it builds brand allegiance, and brand allegiance lowers the transaction costs of substitution choice. Strong brand allegiance can lower the transaction costs of substitution to the point of negating it as allegiance increases. In other words, the more important the label is, the lower the propensity would be for shoppers to make a choice to switch between labels.

Brand allegiance also allows producers more flexibility to charge higher prices than what economic theorists call "the market clearing price." In economic parlance, the market clearing price is the amount that allows producers to recuperate all costs associated with making the product, from start-up costs (also referred to as fixed costs), to average costs associated with direct manufacturing, to average costs associated with promotion and distribution. For consumers, it is assumed that the market clearing price is the threshold point of upper price tolerance. In other words, it is the maximum price they are willing to pay before switching to a cheaper substitute. Building brand allegiance allows retailers to charge a higher price than the one that simply covers production costs. It is the amount of this markup, to put it in fashion retail jargon, which in this case is an economic rent. Traditional fashion retail, as examined by the economic literature founded on the assumptions denoted in Krugman and Helpman (1985), has focused on the amount of such economic rents in relation to brand strength and allegiance as described by the theoretical literature examined in Chapter 2. In everyday fashion retail parlance, such economic rents are described as "the story of the $200 jeans." They cost pennies to produce, but customers are willing to pay $200 because of the designer brand.

Understanding the customer mentality that would allow the extraction of maximum economic rents defines fashion conglomerate profitability. It is all based on consumer-choice dynamics in terms of what factors influence consumers to make a choice, what factors influence the number of choices they face, and what factors consumers evaluate to derive satisfaction from a purchase. For these reasons, consumer choice is one of the most widely studied subfields in marketing and economics. In marketing, the promotion of choice options takes the focus. In economics, choice is put in the context of utility functions, that is, gaging the satisfaction levels of consumers in the context of choosing consumption bundles. Consumption bundles can be described as trade-offs in purchase decisions among amounts of money spent, amounts of money saved, and the features the chosen products offer consumers. Each consumer has a taste-specific consumption bundle (Robson, 2002).*

All consumption bundles are subject to budget constraints (Perloff, 2010). As noted, budget constraints are the function of disposable income and discretionary spending. As suggested, utility is gained not only from the benefits of the consumed products, but also from the satisfaction of making a choice that is high in value, that is, offers significant monetary savings with respect to the evaluation of substitute options. The high value in consumption bundles is contingent not on the direct money spent on the products, but on the money saved. This particular feature of utility, put in the context of choice as one examines elasticities of price in apparel, is at the core of the promotional culture of modern fast fashion. Its advertising strategy is to communicate high value through significant monetary savings.

This strategy has grown of late, as the successful fast-fashion marketing models have shown retailers that customers value low prices more than the historically accepted standard of status. The growth of the strategy is a product of the social factor advances in lifestyle and economics discussed in previous chapters. It is important to understand that the historical assumptions of utility are challenged because the social factors that have defined the demand and supply of apparel products have changed.

The most often studied of these social factors defined the assumptions of fashion theory. These assumptions can be separated into three main pillars: (1) the style pillar—the belief that public tastes dictate style; (2) the status pillar—the perception of luxury items in social contexts; and (3) the comparison pillar—the evaluation of utility compared with other products (Dick and Basu, 1994; Miller et al., 1993; Vigneron and Johnson, 1999).

---

* For a thorough discussion on the evolution of consumption bundles, their relation to choice and price, and that relationship's impact on the elasticity of substitution in the context of the theory of nonsatiation, see Robson (2002).

The extant literature on fashion theory agrees that these pillars offer the generally accepted truths behind the buying behavior of apparel customers. These truths are that style is a reflection of public tastes, that luxury offers utility because it denotes status and success, and that this utility is different because it offers value that is not subject to budget constraints (as most luxury-item buyers value cost savings less than the iconography of the symbol of the brand). However, as Chapters 1 and 2 indicate, it is not clear who dictates style—the public or the industry. In addition, as the evidence of the fast-fashion proliferation of conglomerates such as Zara, H&M, and Forever 21 indicates, they are able to successfully compete with the luxury brands by offering features indistinguishable in similarity. But most importantly, as H&M showed the world, the evaluation of utility against other products' paradigm, as embodied by the "on sale" tactics of traditional apparel retail, have been replaced in terms of both scope and choice. Modern fashion consumers are not most likely to buy an item if it is "on sale," which is the assumption of pillar 3 in fashion theory. They are most likely to buy the item if it is priced low at the outset.

When discussing assumptions of economic theory, it is important to stress that there is a technical difference between "more likely" and "most likely." More likely denotes lower opportunity costs associated with making a wrong choice. In economic theory, an opportunity cost is the cost of the second-best alternative. In plain words, when a customer is evaluating two items, one priced high but more prestigious, the other priced lower, if the customer chooses the first item and pays the premium price, the opportunity cost is the difference between the premium price and the price of the item that is on sale. When economists model price-elasticity opportunity, costs define the level of elasticity. Holding budget constraints equal, the higher the opportunity cost, the higher the probability of choosing the lower-priced item and sacrificing prestige for monetary value. This is the assumption only if all customers value utility in the same way. However, in fashion they do not. Some value prestige, some value design, and all value novelty. Therefore, it is more appropriate to use the term "most likely" when studying the elasticity of fashion sales.

As already noted, in fashion, customers' buying habits are remarkably inelastic (a) because of brand allegiance, (b) because of cultural pressures to remain fashion forward,* (c) because buying is done for pleasure, not utility, and (d) because prices are decreasing in a downward spiral,

---

* This particular fact is explored in depth in Chapter 7.

diminishing the opportunity for rent extracting. Therefore, in a retail market when the opportunities for rent seeking are high, the classification of "more likely" is best suited when modeling consumer behavior, because it denotes that the analytical results of the economic model reveal the amount of economic rent. In retail markets with low rent-seeking possibilities, the classification "most likely" is appropriate because it describes a dichotomous state—buy or not. When prices are so low as not to be included in the evaluation of choice as a substitution attribute (which is the case in fast-fashion retail), buyers decide whether to buy items because of their nonprice attributes. Choice is present, but not in terms of price. In the fast-fashion context, customers choose among many equally attractive items in terms of price. Everything is cheap, and although some items are slightly more expensive, they are more expensive because of product type, not because of better labeling. Even these relatively more expensive items are still cheap enough in the eye of the modern consumer. Therefore, ultimately the decision is to either make a purchase or not. It is not whether to make a purchase, make a "second-best" purchase, a "third-best" purchase (and so forth), or not. This dynamic is imperative to keep in mind when studying sale elasticities in fashion retail. It is related to the embrace of and reliance on impulse-buying promotional tactics by fast-fashion retailers. These tactics are the focus of Chapter 7.

Another very important aspect that affects the elasticities of fashion sales is seasonality, as embodied by the traditional concept of a fashion season. As described in Chapter 1, traditional fashion merchandise, its accessories, and its promotion have been based on eight distinct seasons. Therefore, fashion theorists had focused on studying the behavior of fashion-forward customers with respect to the excitement of seasonality.

Tracking the research on seasonality of fashion purchases in relation to price elasticity, Martinez (2012) notes the historical findings that clothing sales exhibit seasonal fluctuation (Kopp et al., 1989; Allenby et al., 1996; Wagner and Mohktari, 2000). However, the premises of these works on the whole concept of seasonality are unclear. Allenby et al. (1996) follow up on Kopp et al.'s (1989) definition of "season" into a dichotomous concept of "pre-season" (January–March) and "in-season" (April–June and October–December) to examine how the argument posited by Kopp et al. (1989), which states that shoppers exhibit seasonal preferences, can be used as a predictor of fashion sales. The problem is that, as already discussed in Chapter 1 and outlined by Birnbaum (2005), even back in the 1980s and 1990s there would arguably have been such a dichotomous seasonality.

There were eight distinct seasons, broken down from the famous two main seasons—fall and spring—into four seasons denoting the retail dynamics of winter clothing and the retail dynamics of summer clothing. Furthermore, Kopp et al. (1989) use survey data that do not specifically address season questions, but instead ask for demographic information from respondents. This information is then coded by the authors into categories for "fashion enthusiasm" to see how the economic conditions of buyers can be understood in the context of retail stores competing for their business. Not surprisingly, the results suggest that wealthier shoppers are more fashion forward and exhibit higher propensity for chasing the latest fashions that are most aggressively promoted during the hype around fashion week schedules.

Putting this information in the context of price elasticity to study how substitution choices and price fluctuation can be used in a competitive platform would be misleading for several reasons. First, modern retail does not compete on a concept of "pre" and "in" season. It competes in a reality of at least 20 seasons that are equally promoted, without any specific weights. Second, its nature has changed toward consumption promotion that is very different from the historical "fashion-line" merchandising platform of stocking shelves with merchandise once every 2 months and gradually decreasing the initial price through markdowns. Third, the demographics of the main purchasing market have decreased from the traditional adult female shopper to the younger demographic of teenage consumers. All classic fashion retail research takes for granted the assumption that adult women are the main economic demographic of clothing retail. Chapter 7 explains why this is no longer the case. Fourth, and most importantly when it comes to pricing strategy and retail competition in the context of elasticity estimation, the promotion of fashion has changed the commercial model of the fashion industry from branded retail and label centered to impulse purchase driven.

7

The Economics, Demographics, and Ethics of the Low Price Quest

THE INDUSTRIAL PSYCHOLOGY OF IMPULSES

Fast fashion relies on impulse buying. Fashion retail in general has developed a gradual proclivity toward embracing the impulse purchasing model. The reason for this fact is historic and has to do with discretionary spending. Developed in the United States, the impulse-buying retail model was a product of the gradually increasing disposable income and discretionary spending of Americans after World War II. Madhavaram and Laverie (2004) offer a thorough chronology of impulse purchase marketing and track its advent and proliferation as a managerial and academic subject to the 1950s, starting with the seminal definition of impulse-buying behavior by Clover (1950). The author outlined the phenomenon in the natural experimental context of what are known as the unplanned 1948 gas holidays. That year, in three Texas towns, gas shortages forced all businesses to close for the day on two occasions. Clover (1950) interviewed hundreds of local retail managers trying to understand strategies for dealing with lost sales. The results revealed that impulses strongly influence sales, and that these impulses could be stimulated by retail environments, and therefore were subject to in-store experiences.

Since that time, impulse purchasing has been studied across different industrial sectors. It has become accepted as a generalized consumer trait consistent across product categories. The main conclusion is that shoppers will succumb to the impulse of making unplanned purchases while in a retail environment (Jones et al., 2003).

It is important to note the inception of the impulse-buying model, because impulse purchasing was first observed, analyzed, and promoted as a business model in clothing and home good retail environments.

Clover (1950) hypothesized that stores that were relatively more reliant on impulsive customer behavior would be less likely to make up the lost sales from the "gas holiday." The logic is that since their business model already relied on impulse sales tactics, not much could be done to increase sales to make up for a lost day of work. However, the author found that to be true only to an extent. The findings revealed that the loss effect was mitigated by the aggressive action of retail managers in those particular stores to offer strong in-store stimuli. Taking that concept further, Applebaum (1951) noted that impulse behavior was more pronounced in environments where the natural senses were highly stimulated. Applebaum's research was on grocery-store impulse buying, and it was suggested that tactile stimuli, ranging from the smell of the store to samples of free food, provide a strong enticement for consumers to engage in unplanned purchasing.

Undoubtedly, the legacy of such research is at the core of the business models of most modern food retailers, including box stores such as Costco and BJ's Club. These two research works by Clover (1950) and Applebaum (1951) are seminal because they lay the foundations for the understanding that the retail-store environment is the defining element of impulse buying. The two core factors of the impulse-buying model are (1) customer service, that is, the retail management's role in enticing unplanned purchasing, and (2) store atmosphere, that is, decor, promotional aids, smell, tastes, and unadvertised "finds," such as in-store-only sales and product promotions.

Since the 1940s, retail management has developed around these two core tactics of customer experience. Because of this strong legacy of impulse-driven retail, and particularly the importance of in-store unadvertised sales, the fast-fashion model of "no sale" proliferated by H&M* was highly contested at the outset, particularly in America, where it was thought unwise to challenge a culture of impulse consumption that had been gradually fostered for generation after generation. The "sale" component was considered the core foundation of in-store stimuli.

But the difference between the analogy and logic of impulse purchasing management in traditional fashion retail and in fast fashion today is that in the past, impulse buying was considered merely a secondary retail phenomenon (Sheridan et al., 2006). A fairly low percentage of total sales were contingent on unplanned purchases. Today, the business model has been augmented to heavier, if not exclusive, reliance on impulse buying

---

* Refer to Chapters 1 and 2.

(Bruce and Daly, 2006, 2011; Bruce et al., 2004; Byun and Sternquist, 2011; Doyle et al., 2006).

Not all buyers succumb to impulses equally, and fashion retailers rely on certain consumers who are more prone to engage in habitual impulse purchasing as their target base. Johnson and Attmann (2009) posit that these consumers are young women who exhibit higher signs of materialism and neuroticism, denoting a general level of unhappiness. A few studies on fashion marketing manipulation examine the susceptibility of emotionally disenfranchised, mainly young, and mainly female shoppers to engaging in habitual impulse purchasing. This is because of an attempt to fulfill psychological needs that stem from insecurity. However, not all research places a negative spin on impulse promotion. Phau and Lo (2004) explain that "fashion innovators" are more prone to engage in impulse buying, and these are people who are generally happy, well adjusted, excitable, "contemporary," and "liberal." The focus of that particular research is on understanding whether these fashion innovators carry their in-store impulses to the online retail environment, and the findings suggest that the answer is yes. But it is the explanation and tracking of previous findings on fashion innovation that offers important insight into the proliferation of impulse-buying behavior. The conclusion is that impulse purchasing has moved into a realm of social acceptance as the leading pattern of consumer behavior. The implications for management are to accept it, promote it, and improve ways to entice customers to increase the *amount* and *frequency* of *unplanned* purchases.

Many retailers outside the strict fast-fashion model also rely on impulse purchasing. Impulse purchasing behavior has given impetus to the creation of a wide range of suppliers, from big-box stores such as Costco to value retailers such as Marshall's. Although different in their targeted demographics and products, the common feature these retail establishments share is their low prices. Customers are drawn to them in search of savings. Once inside, customers tend to make many unplanned purchases because of value stimulation, that is, seeing surprisingly low prices in unexpected in-store sales promotions. The degree of such value stimulation depends on the store environment and its propensity to create impulses.

With respect to apparel purchases, buying impulses are stimulated on site. Therefore, the store experience is essential (Mattila and Wirtz, 2008). Bridson and Evans (2004) describe the trends in developing different store experiences among competing branded retailers, and track how fashion retailers have oriented their marketing tactics in favor of offering better store environments. The term offered to describe the consistent uniformity of product, service, and environment across all retail stores is *brand orientation*. Bridson and Evans (2004) go as far as to argue that brand orientation is the "secret to fashion advantage," as the title of their paper states. The explanation is that brand orientation is the only aspect of retail that offers an avenue to build uniqueness, because fashion competition lacks the two most often employed competitive attributes of other retail sectors—differentiation and sustainable competitive advantage.

It seems counterintuitive that fashion, an industry that promotes its attributes based on high levels of choice and variety, should be defined by a low level of product differentiation. But the statement has merit and is accepted as a general truth in fashion. The low level of differentiation stems from the fact that trends dictate that certain clothes attributes are fashionable, and all retailers must follow what fashion dictates. The result is clothes sharing common trends and similarities among different retailers. This similarity is at the core of the other competitive problem fashion retailers face—building sustainable competitive advantage. The difficulty in building long-term competitive advantage comes from high aptitude for replication. Technological advances have made it easy for competitors to replicate each other with speed and ease. Brand orientation also has impacts outside the physical retail space, that is, in the online shopping platforms. Park and Lennon (2009) find that the actual store image, as embodied by perceptions of past shopping experiences, positively influences online shopping behavior, making customers more likely to make purchases online.

The store experience is the essential venue for employing marketing tactics to stimulate impulses. Mattila and Wirtz (2008) study what they describe as overstimulation—higher than desired levels of stimulation. They are "higher than desired" because they are expensive for managers to implement. Even with the added costs, such as additional hiring, expensive décor, and theme decorating that changes often, the argument is that it is no longer enough just to stimulate; a successful retailer must overstimulate in order to increase the incidence of unplanned purchases. Adding to this body of research, Park et al. (2012) focus on clothing impulse purchasing online, and also stress the importance of strong stimulation in the online environment. The authors found that impulse buying was positively associated with visual stimulation of sensory attributes beyond mere photographic images. In other words, simple pictures presented in a catalog-like style only take stimulation so far. However, video clips, music, and filmed advertising segments provide that crucial extra stimulation, or as Mattial and Wirtz (2008) put it, overstimulation.

In the United States in particular, overstimulation has increased not only in clothes retail but in most retail merchandising (Schor, 1998, 2004). Cultural factors and decades of general economic growth have enabled the continuous testing of different tactics in promotion. The result is a variety of marketing models separated by segment and intensity. Among them, impulse-buying is accepted by professionals as the most important, and generating impulse-buying volume is most successfully achieved through overstimulation (Kalla and Arora, 2011). The tactic has been particularly successful at targeting younger and younger people (Bakan, 2012).

In an acclaimed and integrated body of work on consumption factors, Schor (1998, 2004, 2008) offers a comprehensive analysis of the reasons why Americans across several generations have successfully succumbed to their buying impulses. The reasons range from the competitive nature of the nation, which impacts the desire not to be left behind in possessions, to the low disposable income brackets discussed previously in Chapter 6, which allow people to expand their consumption bundles, to the creative and increasingly emotional impact of overstimulation marketing. The result is a culture of consumption that is built at a very early age.

Increasingly, in all industries, much attention is given to advertising aimed at younger and younger consumers. In fashion, this trend has redefined market demographic dynamics. It has been very successful for several reasons. One is that American children and teens enjoy significant allowances of their own, which most prefer to spend on clothes (Dotson and Hyatt, 2005*; Furnham, 1999; Moses, 2000). Moses (2000) shows that this American consumption trait is successfully exported around the world. Globally, teens are driving significant retail commerce, and it is based on allowance amounts that have gradually and steadily increased in the past three decades.

Globalization in production and global brand advertising strategies (as explained previously in Chapters 1 and 2, in relation to the average consumer) have impacted child consumption patterns toward augmentation.

---

* In addition to this point, Dotson and Hyatt (2005) explain that these young children have significant influence over the entire consumption pattern of their families. This is a unique modern cultural development, because it has changed the direction of purchasing decision-making from parent to child, to child to parent. Furnham (1999) notes that children as young as 9 years old have significant amounts of money in their "piggy banks," which marketers aggressively target.

The result is that children all over the world either make direct purchases on their own or successfully convince their families to buy them an ever-increasing amount and variety of goods. Another reason why younger shoppers are becoming the defining demographic of fashion sales in the United States in particular is that many young Americans start to earn their own money early in life. They are encouraged to seek employment by both their families and peers, because working is seen as an important character-building virtue (Csikszentmihalyi and Schneider, 2000; Loughlin and Barling, 2001). This reason creates a whole stratum of disposable income in shoppers as young as 12 years old. The majority of this income is devoted to apparel. The main reason apparel sales are growing so fast in the youth markets is that the price of adolescent apparel has decreased to such a low level that no other child or adolescent product, except for fast food, can compare.

THE MORAL UNIVERSE OF NONWAGE EARNERS

When sweaters and T-shirts cost less than a sandwich, they fall into discretionary spending brackets that children can afford. The low prices also enable parents to indulge the clothing wishes of their children, or at the very least, to object less to the expense associated with keeping their young stylish. This is the reason why stores such as Walmart and Target compete to become leaders in the quickly growing fast-fashion industry, which is a new retail venue in fashion retailing. Buying stylish clothing in value chain stores such as Target, Kmart, and Walmart would have been considered laughable a decade ago. In the 1990s, there were whole lines of jokes about Kmart clothing. The worst thing a teen could do was admit that an outfit came from Target. But fast fashion has changed that mentality. The ability to manufacture stylish clothes and retail them at cheap prices has attracted designers who create lines for value retailers in search of wider audiences. The value retailers promote those product lines aggressively, but unlike the branded retailers who also rely on the same designers, value retailers promote the low prices above all. This tactic has worked and has changed the negative attitude toward value retailers. Today, they continue to sell cheap garments that can be used and discarded with ease, but nowadays the quality and attributes of these garments are often indistinguishable from the designer-brand options. Modern budget-conscious teenagers love the concept of very inexpensive clothing and display pride in the fact that their stylish outfits cost very little. The low prices allow them to shift styles with speed and on a low budget (Bae and May-Plumlee, 2005; Barn and Lea-Greenway, 2006; Bruce and Daly, 2004, 2006). Understanding that the marginal growth lies with the young has been essential in the promotion of fast fashion. For these reasons, the entire fashion demographic market has been steered toward youth, as has been its entire marketing platform.

Marketing to teenagers separately from adults has been in use in the West since 1941 (Quart, 2008). From then on, the size of young market demographics in relation to older brackets has increased in terms of both value and incidence of overall purchases. What is new and, as already discussed, somewhat surprising to marketing professionals is that unlike their adult counterparts, teens, tweens, and children share a greater degree of taste similarity internationally, particularly in branded products (Wee, 1999). There are many reasons. One of the main reasons is that "teenagers are a target for considerable marketing investment as fashion companies attempt to gain long-term loyalty from potentially affluent emerging adults" (Cassidy and van Schijndel, 2011, p. 164).

The way that fashion has been marketed to teens has also changed. In the 1980s and 1990s, teen fashion was promoted much like regular fashion, through the pages of teen fashion magazines such as *Seventeen* in the United States and *Jackie* in the United Kingdom (no longer in business). The main focus back then was to promote budget buys. The editorials stressed messages of frugality and utility because of the general understanding of the budget limitations of young shoppers. Quart (2008) claims that the result was a majority of clumsy fashion spreads featuring downmarket items. The teen market was seen as a "down market"—one in which marked-down clothes were sold. Fast fashion changed that legacy.

The demographic target markets have gradually become younger because of major changes in several perfectly confluent factors: (1) technology, (2) prices, (3) celebrity, and (4) parenting. Among these four factors, as discussed in Chapter 2, technological innovation is undoubtedly the foundation. In addition to improvements in fabric manufacturing, technological improvements profoundly impacted fashion advertising. The main outcome can be summed up in the advent of *new media*—Internet and cellular platforms that are accessed through "smart" devices.

New media brought about immediacy of communication, broad contexts of message promotion, low transaction costs of reaching consumers, and an ability to engage the consumer emotionally. Furthermore, through varied communication channels such as blogs and other virtual input options, technological innovation has provided a voice to consumers in a direct and peer-related way. Before the proliferation of the Internet and social media, traditional fashion promotion relied on in-store customer input, or, as tracked by the volumes of academic research examined during the writing of this book, surveys. Surveys were the core way of gathering information on customer taste and preferences. Although useful in gathering insight, surveys cannot be compared to the information exchange options in social media, because these options show not only how customers react to the questions of a survey, but also how they react to input from other customers, as opinions are formed in open forums. Customers get to connect to other shoppers, and they seek out similarities with others in search of validation. Discourse ensues in the open blogosphere, where information is multidirectional and multifaceted. In other words, consumers exchange information with producers, promoters, and other consumers all in one place. The discussions vary from garment features, to quality, to the artistic merits of the advertising campaigns, to the political implications of the social messages in these campaigns. All these topics are discussed because they all influence the creation of style and relate to the advent, promotion, and proliferation of trends. In such a rich environment of information exchange, fashion magazines no longer dictate primary and secondary markets.

The second main factor due to which the fashion demographic markets are converging toward youth is price. The whole reason for writing this book was to examine the pricing dynamics of an industry that has its history in pricing models based on rent seeking and rent extracting through markups defined by brand strength. The industry changed when a few bold and undoubtedly well-meaning retailers, such as H&M and Zara, decided to compete against the rent-protecting platform of the branded retailers. They embraced the technological advances in production that offered an ability to significantly decrease production costs, and decided to pass the savings directly to the consumers while increasing the core style and trend features of apparel products. Branded retailers stayed true to the rent-seeking model of brand allegiance, and that is the main difference why young consumers became the fastest-growing apparel target market. Their lower budgets could not afford *many* branded purchases. However, these same lower budgets allowed *many* fast-fashion substitute options. Gradually, as the size of the younger demographic segments increased, a separate age-specific commerce in teen fashion emerged. Then, the branded retailers started to look toward younger customers. Over time, everyone has either lowered their prices or offered youth-led fast-fashion-priced product lines. Famously, in 2012, Versace launched its fast-fashion collection for H&M to unprecedented commercial success that continues to this day. Versace for H&M is a global success, promoted by the most famous celebrities in the world, such as Madonna and David Beckham.

The third factor is related to the abovementioned celebrities. It is the change of apparel spokesmodels. The traditional fashion model—anonymous and elusive, although aspiring and admired—is replaced today by a celebrity. Celebrity models spearhead advertising campaigns for teenagers and children because young people are much more influenced by celebrities than mature consumers would be (Doss, 2011). This is the case because, as Okonkwo (2007) explains, celebrities wield significantly increasing influence in the main cultural facets of society—arts, music, movies and television, sports, culture, and politics. Young people feel a connection to celebrities because of what the celebrities do, not just how they look. Looks were the main attractive feature of traditional fashion models. Their beauty sold the products. Today it is not only the celebrities' beauty (for you'd be hard-pressed to find an unattractive celebrity model) but also their professional skill and, as will be discussed under "The Elasticity of Dirty Consumption," their carefully managed legacies of social activism.

Celebrity endorsement has a social activism component because it does not start and end with a photo. Celebrity image is part of a celebrity's carefully cultivated legacy. When a celebrity has a reputation for social activism, community engagement, or even religious activism, that reputation translates to the brands he or she promotes. This also works the other way around. Doss (2011) calls it *brand transference*—a process describing the social impact of the brands on the image of the celebrity. Celebrities embrace brands that they believe are socially responsible, and in return become known as activists themselves. Furthermore, celebrities carefully choose a package of brands to promote (Um, 2008). For these reasons, teenagers respond to the promotions of celebrity spokesmodels with more enthusiasm and greater allegiance.

The fourth factor that contributes to the importance of younger fashion demographics in the overall sales volume of the industry is a culture of parenting that is merging toward equality. Today, social factors have begun to erase the traditional hierarchy between parent and child.

In developed nations, where these dynamics have had the strongest impact on demographic market changes, the difference in authority between teens and their parents has become particularly small because parents try to be friends with their children (Bakan, 2012; Bornstein and Bradley, 2014*). With respect to fashion purchases, this cultural trait has two main implications. One is that parents are less likely to refuse the fashion whims of their children. The second is that the children are becoming the trendsetters for their families. In this role, they drive the apparel and related good purchases of households. Bellman et al. (2009) explain that today millennials wield great influence over the buying choices of older consumers. It is a claim that goes contrary to traditional logic, but is much supported in the consumer demographic literature on the proliferation of fast fashion. Phau and Lo (2004) posit that teenagers are, on the whole, the new fashion innovators. The authors explain that such innovators dictate the style choices of their social and familial circles. Their style choices and their behavior are emulated by younger shoppers, that is, children, and older shoppers, that is, their parents, who see them as style decision makers. They are the source of information of what is fashionable for older consumers who wish to stay current and fashion forward.

There has been great interest in analyzing teenage fashion consumption. Insightful points come from the literature on fashion merchandising and also from the sociological fields that link commercial behavior to socialization. For example, Colucci and Scarpi (2013) show that marketers today target generation Y purchasing patterns because young people shop for fun, not for utility. Park et al. (2006) explain that the behavior of shopping for fun has become the focus of a separate industrial psychology research body, appropriately dubbed "hedonistic consumption tendencies."

This is a reality rooted in a "hanging out at the mall" culture (Matthews et al., 2000). Since the 1980s, teens in the Western world have instituted a playground mentality, so to speak, of Friday-night mall outings. This culture is embodied in the phenomenon of "mall rats," as it became known in the 1980s. In her seminal work *No Logo* (1999), Naomi Klein explains the culture and relates it to the promotion of consumerism. That particular

---

* Bornstein and Bradley (2014) offer an integrated volume of works from the leading experts in parenting sociology. The individual chapters focus on specific issues of modern-day parenting in a global context, from historical factors to cultural and economic change. Each contributor in the volume notes in one way or another that parents today face pressures to discipline softly, not to refuse children's wishes (or to do so minimally), and to provide more and more rewards as incentives in child behavior modification.

work analyzes the factors that built this culture of consumerism in order to put it in the context of global social justice issues. Klein's main points are that the commercial interests of multinational corporations are the main drivers behind consumerism, and their quest for profit is behind serious human rights abuses. Some of the research examined in Chapters 5 and 6 of this book offers support for Klein's claims.

Since that time period, the "mall rat" culture, mainly associated with middle-class American teenagers in the 1980s, has proliferated to all societal strata in America and other industrialized nations with similar retail environments (Parker et al., 2004; Spilková and Radová, 2011).* Today, around the world, it is in malls or other such industrial retail and entertainment parks that young people socialize in their free time (Sit et al., 2003).

In the United States, where mall socialization began, it is typical as children grow to tweens, teens, and young adults that their parents allow them to engage in unsupervised social mall outings as a testament to emancipation. During these instances, the young people mainly shop for clothes, shoes, food, and, to a lesser degree, cosmetics and personal electronics. The typical teenager goes to the mall with friends in an institutionalized social convention to spend a weekly allowance on entertainment (Bakewell and Mitchell, 2003; Haytko and Baker, 2004; Matthews et al., 2000). Most of this "entertainment" is in the pleasure of shopping for clothes (Taylor and Cosenza, 2002). This means that buying clothes has become an institutionalized weekly habit. Bakewell and Mitchell (2003) examine research on how much time American teens spend at the mall and estimate that on a weekly basis it is around 15 hours, broken into at least two visits.

The outcome of these facts is reflected in Bellman et al. (2009), who posit that in America the millennial customer market for fashion and related products is estimated to be over $1 trillion annually. Kathy Grannis of the American National Retail Federation estimates that in 2014, over $900 million of that $1 trillion was spent during the back-to-school shopping

---

* Spilková and Radová (2011) explain that this American-based behavior is extremely well embraced by teens in the fast-paced industrializing metropolitan centers of Eastern Europe. Other researchers, such as Parker et al. (2004), have shown the same dynamic to be true in Asia, and the majority of fast-fashion retail research, such as Bruce and Daly (2004, 2006) is based on UK retail. It is very telling that the concept is so well emulated in Eastern Europe, because this region is still relatively less developed due to the legacy of communism and the transitional decade of the 1990s. The implications are that people there have less money to spend in such indulgent shopping experiences, and yet the allure of Western-style behavior fuels the allocation of very limited resources to this type of consumer behavior.

period of late summer.* From it, the average household spent about $700 per child, of which less than $300 was on nonapparel items such as consumer electronics and actual school supplies. The majority of the expense was devoted to new clothes.

In the context of previously noted research, which estimates the entire annual global market for clothes to be around $1.5 trillion, these metrics suggest that the habitual shopping of American youth comprises the majority of global fashion profit. It must be noted that the $1 trillion estimate Bellman et al. (2009) (and others) offer is based on calculations from the US Census bureau that are at least 10 years old. With this fact in mind, the metric offers an additional insight: specifically, that the real value of global fashion commerce is much higher. As already discussed in Chapter 6, this commerce includes the trade in used clothing, which is estimated at around $500 billion. Also, if one takes into account the most current data points presented in Chapter 1 on store openings in emerging markets, that is, that Inditex and Adidas alone admit to having opened an average of one new store a day in the developing world for the past few years, it is highly unlikely that the magnitude of such growth is worth only (approximately) $300 billion a year.† However, even with such a lack of precision in the dollar value of the entire industry, if indeed over two-thirds of profits depend on American teenagers, a third and extremely important point emerges, which has defining implications for fashion economics theory. This point is that the market segment that drives total garment sales is overwhelmingly comprised of people who either are nonwage earners or earn little compared with the income brackets of older generations.

The elasticity of consumption of nonwage earners is very important to consider, because nonwage earners enjoy an insulating buffer from economic pressures. Therefore, their discretionary spending brackets remain relatively unaltered by economic downturns. This is the case because their discretionary income is allocated in the context of family expenses, and in that context it is relatively small. Compared with large expenses of an "economic unit," as the Bureau of Economic Analysis defines an average American household, teenage spending money amounts are small in relation to mortgage, car, insurance, food, and debt payments. When economic conditions deteriorate for a household, the wage earners (parents)

---

* See Grannis (2014). Grannis also posits that the average household spends upward of $300 per child, and about $23 of that $700 is from the child's "own" money. This ratio is a testament to the high reliance of American youth on allowance money.
† Refer to estimates in Chapters 3 and 4.

tend to engage in savings in large discretionary bracket categories. These bracket categories are delineated for high-priced items or leisure expenses such as vacations, new cars, or other durable goods, that is, large household items such as furniture, refrigerators, or washers and driers, which have a life span of 5 years or more.*

In economic terms, the consumption bundles of nonwage earners are less subject to budget constraints because budgetary allocations are predefined at relatively low levels and are overwhelmingly delineated toward specific type of purchases. In plain language, nonwage earners get most of their income in the form of entitlements (allowances) that are specifically designated for a specific purpose—buying something at the mall. Therefore, consumption bundles are uniform. For example, few teenagers would choose not to buy clothes with their friends around them but to save the money in order to allocate it to other future, more expensive purchases, such as a car or a bike or a computer. In today's world, these types of purchases are delineated differently in household consumption bundles.

Additional cultural reasons exist that lower the elasticity of consumption of nonwage earners. These reasons center on the socialized acceptance of impulse purchasing. Impulse purchasing defies rationality because utility is based on gaining value that has an unclear monetary equivalent. Therefore, it is based on the strength of feeling and emotion. The stronger the emotion, the lower the elasticity of consumption.

THE ELASTICITY OF DIRTY CONSUMPTION

Impulse purchasing is global. Lee and Kacen (2008) show how pervasive the behavior is across cultures. The authors' focus is on understanding cultural factors that entice customers not only to engage in unplanned purchasing, but also to repeat that behavior. An interesting observation is that in collectivist cultures, as embodied in their research by samples from Malaysia and Singapore, customers tend to gain higher levels of satisfaction from impulse buying when others are present. The conclusion is that

---

* The traditional definition of a durable good is a product with a life cycle of 5–10 years. Durable goods are replaced less often and therefore cost more than nondurable, or fast-moving, consumer goods.

peer accord increases impulse purchasing because of the affirmation of a social circle. The implications from this conclusion are very important, because they reveal certain facts about the societal embrace of impulse purchasing in the developing world.

As noted, impulse purchasing was invented, so to speak, by Western marketers. Its foundations were laid in America, where marketing models at the retail level have fine-tuned the promotional tactics of enticement that increase impulse purchasing. Impulse purchasing has been so prolific in impacting consumer behavior that it has become the defining way in which Americans shop. This is the case because all purchase excursions, planned or spontaneous, tend to include the buying of items customers did not plan on purchasing. Impulse purchasing is behind the problem of irresponsible spending that has resulted in such cultural phenomena as hoarding and credit-card debt issues that increasingly lead to bankruptcy (Zhu, 2011). Irresponsible spending is a problem in the West because of the high discretionary spending brackets previously discussed.

In light of this information, Lee and Kacen's (2008) study helps explain why the proliferation of fast-fashion retail is growing in the developing world at such accelerated rates. It is the combination of the fact that collectivist cultures are more prone to engage in habitual social behavior and the fact that Western brands have a special allure in the developing world. They bring status. The status Western products carry is very important in developing countries, and is a main retail stimulant. It is imperative to relate this fact to the economic history of such nations, because that history has built a culture that values saving rather than spending. And yet, customers there are successfully lured to break with tradition and shop with indulgence.

In developing countries, particularly in Asia, Southeast Asia, and Eastern Europe, a cultural legacy of abstemiousness and frugality, particularly embraced by former communist governments, has formed a consumer culture of thoughtful and planned purchasing. As a result, much research on consumer behavior notes the high rate of savings in China, for example, and relates this fact to the history of poverty that has contributed to an ascetic mentality whereby savings are revered and profligacy is considered a vice (McKinnon and Schnabl, 2003; Nabar, 2011; Wei and Zhang, 2011; Yang et al., 2011). An additional factor that contributes to the relatively high savings rate of Chinese and other nationals in countries with communist pasts is the historical lack of product variety, which was (and in some cases still is) due to planned economic production management. A third important aspect to consider is the fact that low discretionary spending further contributes to lower propensity to engage in unplanned purchasing. In light of these facts, Lee and Kacen's (2008) findings have important implications in understanding customer behavior in adapting to external platforms of retail and responding to foreign advertising tactics. The findings show that impulse purchases increase as a function of communal behavior. In other words, young people engage in these purchases in groups. Therefore, the approbation of this behavioral pattern extends past individual consumers. The implication is that even in societies where savings are valued over spending, stimulating tactics used to entice customers to act on impulse can be successful if the decision-making process to buy or to save occurs in groups.

This dynamic is not unique to Asian and other collectivist cultures. In America, which is considered an individualistic culture, teenagers also congregate in groups in malls. Their socialization relies on mall outings because of several specific features of American life that limit the options for young people to interact. Because of the urban sprawl that has defined housing development in the United States since World War II, modern American children and teenagers, just like their parents, have significant difficulties in being with their peers outside school. It is seldom that parents allow young people to leave the house unaccompanied by adults, and even if they do, in the average suburban neighborhood the average American teenager does not have many other teenagers to meet. In cases where there are few young people living within a safe walking distance of each other, there is still a lack of noncommercial social space for them to interact. If public space options are available, most are designated playgrounds that are most appropriate for very young children. As children grow older, it is culturally awkward to congregate on playgrounds occupied by toddlers and their parents. For these reasons, young people in America, as well as their parents, embrace the benefits of socialized group outings to the mall. Unlike neighborhood social interactions, the mall environment offers a higher degree of freedom to parents, both in terms of supervision and access. Geographic distance and transportation are no longer barriers that limit the number of choices of friends and playmates. Granted, an adult—or in a very few cases, in large cities, a reliable and safe public transportation system—must aid in the transportation of teenagers to malls, but this is a chore adults willingly undertake, because it allows them a break from direct supervision. Malls are considered fairly safe places because they are well policed. Therefore, parents welcome the opportunity to leave their children unattended there for a few hours. But this convenience comes at a cost; they must give their children money to spend at the mall.

As tracked by the literature on mall culture retail noted earlier in this chapter, the amount of that money allowance is the essential economic fuel that powers modern-day apparel sales in America. Therefore, it merits in-depth analysis to understand its variability, stability, and elasticity. Its elasticity directly impacts the sales elasticity of the industry. An additional reason to study it is because it is an example of limited discretionary spending that has implications for utility functions of two demographics. One is the utility individual shoppers can derive through its allocation—the happiness children gain from using their allowance to acquire new clothes. The other is the utility of the original budget allocator—the parental unit. This utility is gaged by the amount of allowance and denotes the importance parents place on responding to the economic demands of their children.

When buyers have limited discretionary spending amounts, they have lower latitude for change in consumption patterns. They cannot easily increase their consumption. They can decrease it, but such a decrease is unlikely because of possible loss of utility in terms of lowering an accepted standard of living.

Different demographics approach the spending of limited discretionary income in different value-laden decision processes. Older shoppers with limited funds show a higher propensity toward savings. In comparison, younger shoppers with limited funds show a greater propensity toward spending. This is the case because they place greater importance on using limited funds in the creation of their image than older shoppers. This image-building self-definition is a product of increasing socialization needs (Dotson and Hyatt, 2005*; Simpson et al., 1998). Therefore, as Colucci and Scarpi (2013) explain, fashion retailers targeting young shoppers can decrease the elasticity of marginal sales.

For the industry at large, as already noted in Chapter 6, fashion sales are largely inelastic. The opportunity to decrease their elasticity further by changing the targeting of demographic markets toward consumers with the lowest discretionary spending brackets is counterintuitive to traditional economic assumptions. It is counterintuitive because the nonsatiation rule of utility theory states that increasing, not decreasing,

---

* In addition to traditional retail purchases, Dotson and Hyatt (2005) explain that socialization shopping, to put it this way, has also proliferated the online shopping habits of young consumers.

discretionary spending is the causal factor of increasing marginal sales (Debreu, 1987). The logic is that as income grows, consumers demand an ever-increasing amount of goods and services. Therefore, marketing and sales tactics are best aimed at consumers with large and growing discretionary spending brackets. However, in fashion retail, the opposite targeting tactic has proven to be the successful business model. Yarrow and O'Donnell (2009) explain that this targeting of younger and younger consumers has been so successful as to change the entire demographic makeup of garment retail.

The global fashion market is now based on the shopping habits of teenagers. For specific cultural reasons, today, in the United States alone, teenagers account for close to 50% of the entire clothing and accessories market. They spend five times as much on apparel as their parents did a generation ago (Haytko and Baker, 2004; Quart, 2008).

In the developing world, the youth market is larger because of "youth bulges" that have resulted from improved public health realities, lower infant mortality rates, and fairly high fertility rates in comparison with Western nations (Hart et al., 2004; Urdal, 2006). A youth bulge means that there are more children and teenagers in a society than adults. Birnbaum (2008) estimated that there are more tweens in China who come from social strata with discretionary spending brackets that allow them to be considered viable fashion consumers than the entire retail customer base of the United States. Supporting evidence that teenagers are an important market demographic not only in China, but in Asia's rapidly growing economies, is offered by Dinesh (2012) and Nilkant (2014), who show that teenagers in India are becoming the core fashion demographic, because discretionary spending brackets among them are growing the fastest. This growth is an outcome of the general unprecedented economic growth in the developing world that has occurred since the early 2000s.

Almost exclusively, modern-day economic development research discusses that growth.* A good example of its magnitude comes from a 2014 story in *The Economist Magazine* based on estimates of economists at Citigroup. Between 1995 and 2007, 59% of global economic growth came from the developed world. Between 2010 and 2013, over 70% of world economic growth came from the developing world (*The Economist*, 2014). This reality is behind the most important aspect of modern-day global

---

* For a thorough and integrated summary of the whole field of modern-day emerging-market growth, see Anguelov (2014, 2015), Hassett (2012), Kaplan (2011), and Zakaria (2011).

retailing. For firms that sell products internationally, marginal sales volumes in developing world markets are stronger for most products in relation to marginal sales volumes in developed world markets (Dunning, 2013; Melitz and Trefler, 2012; Taylor and Thrift, 2012). Apparel sale dynamics are particularly defined by this economic shift and, as explained in Chapters 1 and 2, dictate market expansion strategies for all MNCs that sell fashion and related products. For example, in 2011, Gap announced the gradual closing of 189 of its American stores while expanding in China and Hong Kong by opening 35 new locations (Mattioli and Hudson, 2011).

Furthermore, based on market equalization arguments, future growth-rate potential is higher in emerging markets. This is why retailers are expanding there through not only production, but also proliferation of capital assets, such as building new stores. The growth is due to the liberalizing market change policies implemented in the past two decades, as tracked in Chapters 4 and 5. These market liberalization policies had the goal of increasing the ease of trade. This ease is expected to stimulate capital formation in both emerging and mature markets because of factor/price equalization dynamics.

The assumptions of liberalizing markets state that under factor/price equalization efficiency, as barriers to market participation decrease and markets expand, they expand in scope, while increased competition leads to the equalization of market activity in terms of rates (Balassa, 2012; Jovanovic, 2014). This means that consumption functions, subject to budget constraints, equalize. In plain language, as the economies of the new emerging market participants grow, so does the spending of their citizens.

This growth is fairly recent. Therefore, as a whole, the modern-day global market is unevenly balanced. Most of the wealth is still concentrated in the developed world. The imbalance is such that according to the World Bank, even with the strong marginal economic growth in the developing world,[*] to this day 80% of the world's population live in nations with gross national income (GNI) per capita that would classify them as developing. The World Bank draws that income line at $12,000 a year. Furthermore, over 50% of the world's population live in extremely underdeveloped nations, where their livelihoods are sustained on less than $2.5 a day, which is the official global threshold for dire poverty.[†]

---

[*] Marginal economic growth at the country level refers to growth from year to year.
[†] Dire poverty denotes living conditions hazardous to a person's health in terms of malnutrition, proper hydration, and access to basic health care. Data estimates come from the World Bank database World Development Indicators accessible at worldbank.org.

In unevenly balanced markets, individual consumption functions move toward convergence. As poorer nations climb the industrialization ladder, their consumers—those with lower consumption functions—gradually increase their consumption rates to match the average consumption rate of the overall global market population. The opposite does not occur. The higher-consuming segments of the market population, if one considers the global market as one entity, that is, developed nation consumers, do not lower their rates of consumption toward downward equalization. This is the case because of the nonsatiation assumption of utility theory (Debreu, 1987) and also because increasing marginal consumption is stimulated by economic policies in both developed and developing nations. The foundation of human understanding of economic prosperity growth lies in the concept of increasing marginal consumption.

The nonsatiation assumption is behind the core definition of consumer theory. It describes the fact that customers will choose to increase their consumption bundles in both size and scope toward maximization under their individual budget constraints. This means that all world consumers, in both the developed and the developing world, are predisposed to marginal increases in their consumption patterns. However, the rate of these increases is different, because, given the fact that all consumers will choose to consume more, those with faster-growing discretionary incomes will be able to enjoy higher rates of marginal consumption. In other words, people in the developing world will be able to increase the amount of purchases they make more than their developed-world counterparts. The previously discussed evidence from the apparel consumption of Asian teenagers indicates that such an increase is indeed occurring in fashion sales.

Because of budget constraints in terms of discretionary spending amounts, the nonsatiation assumption only holds true when significant economic growth is present. For nonsatiation factors to lead to global market equalization in terms of consumption, the economic growth must impact wealth creation in the general population with respect to increasing its purchasing power. In that case, consumers with low consumption functions will keep on gradually augmenting them until they equalize with consumers with large consumption functions (Adams, 2013). In applied terms, Dresner (2009) explains the process by offering examples of comparative consumption. According to his estimates, in the 1990s, the average person in North America consumed over twenty times more goods and services than the average person in China and India, and as much as seventy times more than the average person in Bangladesh. These metrics are given in the context of sustainability, warning that as equalization of consumption increases, its ecological impacts could be very significant. However, in a general context, these numbers, as well as other research on rapid market and economic growth in the developing world with respect to consumption rates (Adams, 2013; Hassett, 2012; Kaplan, 2011; Weinstein, 2005; Wolf, 2004; Zakaria, 2011), suggest that future global growth rates there will be higher. This is why fashion retailers such as Gap are opening stores in the developing world while closing stores in the developed world.

Still, the discretionary spending of Western teenagers is not to be underestimated. It should be studied, because it sets the example that emerging-market teenagers emulate. Both in terms of the lifestyle that they observe and try to mimic—with respect to the amounts of clothing, accessories, and cosmetics Western teens buy—and in terms of trendsetting, the literature on cultural imperialism notes that young people in the developing world strive to imitate the lifestyle of their peers in the developed world (Barker, 1999; Chua, 2003*). They are encouraged by the aggressive marketing and promotional tactics of global brands, which to this day exclusively originate in the developed world. The owners of these brands showcase the glamor of Western daily life in their advertising, and young people respond well to it.

The main reasons why this is the case are explained in Chapter 2, but it is important to stress at this point that chief among them is a feeling of inadequacy in the context of global citizenry, which stems from the fact that one is born and lives in a poor country. Very astutely, Wolf (2004) explains how feeling like a second-class global citizen permeates the behavior of people in the developing world in several ways. Among the main traits is a heightened desire to showcase status that can only come from the possession of Western objects. The status implies a certain level of elevation above local economic limitations, because the possession of foreign objects is seen as a sign of international sophistication. Clothes are the most emblematic of those objects. In that respect, as noted in the research by Strizhakova et al. (2008), in emerging markets, global brands are embraced as vehicles for connectedness to a global citizenship mentality. They not only indicate status but also provide an allusion to the

---

* This point is particularly well covered in Chua (2003), and is based on Singapore teenagers' impressions of Western lifestyle during the proliferation of global fashion brands in the 1990s.

emulation of a lifestyle that is richer and for some, more sophisticated. Brand managers promote that lifestyle aggressively and most directly through examples of how young people in America (the most emulated of Western cultures) live and spend.

American teenagers enjoy a level of economic wealth that is unmatched by any other youth demographic in the world. Analyzing this fact, Quart (2008) states that in the United States alone, teens have upward of $155 billion in discretionary spending—the highest concentration of such funds in relation to any other nation. This estimate has to do with their "own" money. However, when it comes to clothing, shoes, cosmetics, electronics, and home product purchases, the funds largely do not come from a teen's own discretionary spending, but from his or her parents. Parker et al. (2007) explain that teenagers spend the majority of their own money on fast food. The money spent on clothes comes largely from their allowances. As an example, to refer back to the data from the American National Retail Federation analyzed by Grannis (2014), of the $700 per child spent on back-to-school shopping, only $23 was from a child's "own" money. This fact is behind the discrepancy in data points between Grannis (2014), who posits that children account for close to a trillion dollars' worth of apparel-related commerce in just one seasonal shopping venture, and Quart (2008), whose estimate of $155 billion for an entire year's worth of apparel commerce is so much lower.

The truth is somewhere higher than the Grannis data, because if $900 million is spent on apparel just in the back-to-school season, one can imagine how much more is spent during the previously discussed weekly shopping trips to the mall. It is during these shopping trips that habitual consumption patterns in young people form. This is also where their parents' shopping habits are influenced, because of two main factors. One is that it is the parents who fund these shopping trips as an institutionalized parenting tactic for the proper socialization of their children. The other is that the parents also indulge their own buying impulses and respond to pressures from their children to remain fashion forward. Therefore, the funds behind these weekly (and, as some research indicated, biweekly) mall outings of young Americans become inelastic to downward economic pressures. The main reason is that habitual spending gets incorporated into family budgets. In other words, parents set an allowance amount, which on the whole does not vary.* This fact provides

---

* Mainly because of the socialization needs it fills.

a very important insight into the lack of elasticity of fashion sales. It is because, on the whole, fashion sales are purchased with inelastic income, and inelastic income is associated with inelastic spending.

In America, teenage income comes from two main sources—one elastic and skewable, the other inelastic. The elastic source is teens' own earned income. In the United States, it has been found that teenagers earn among the highest amounts of their own money, in relation to their global counterparts (Parker et al., 2004, 2007). This is because of the economic makeup of America, where many jobs are suitable for part-time employment, and also because of an entrepreneurial culture that encourages earning behavior (Csikszentmihalyi and Schneider, 2000; Schor, 2008). Still, this income depends on teens' individual part-time job options and availability of transportation. Therefore, this income is skewed toward older teenagers, who have more options due to their ability and capability to drive. But, as so many marketing researchers have analyzed, the age brackets are shrinking to younger, that is, nondriving, teens, and also to tweens who are below the legally employable age of 14 (Cody, 2012; Comstock and Scharrer, 2010; Elliott and Leonard, 2004; Mitchell and Reid-Walsh, 2005). This fact is a testament to the reality that these young people get their discretionary income from their parents or guardians.

In fashion, the tactic pioneered in America to target tweens has been successfully exported globally. Andersen et al. (2008) track the global proliferation of tween clothing advertising and compare how tween retail fares in nations with different social and economic characteristics. The similarities are significant with respect to the amount of money spent.

Not only the American, but the global culture of teenage discretionary spending is based on allowances. In addition to the fact that allowances are inelastic, an important thing to remember is that their amounts tend to equalize among individuals because of socialization information flows. In other words, teenagers in social circles have similar amounts of allowances because of information sharing and pressure from teens to demand that their own allowances are similar to those of their friends. Even if resistant, parents tend to relent and use allowance amounts as reward mechanisms to induce behavioral compliance.

The interplay of continuous demand for monetary reward is cleverly embodied by two terms in the work of Quart (2008): "the gate keeper"—the parent figure who guards the money, and the "nagger"—the child/tween/teen who is asking for marginal monetary increases. For these reasons, fashion and other fast-moving consumer good marketers know how profitable it is to base a retail business model on the discretionary spending of young nonwage earners. In addition, such a strategy offers the possibility to lower the elasticity of sales that extends past current time periods.

When business models are based on young buyers, they offer an opportunity to build brand allegiances in adult shoppers. This is the case because when young consumers enter their prime spending years—25–35—their brand allegiances and shopping habits are already formed (Yarrow and O'Donnell, 2009). Therefore, they will have a lower propensity to engage in substitution, all things being equal. In economics, the disclaimer "all things being equal" means that the realities defining the general societal truths behind the assumptions of the economic theory hold. With respect to the above-examined brand dynamics in economic competition among fashion retailers, the core concepts that must remain equal, so to speak, in order for future sales to be inelastic are: (1) young consumers' favored brands continue to be options and have not disappeared or been crowded out of the market, (2) the brands manage to maintain the same credibility among consumers, and (3) the brands offer the right blend of diversi-fication as the shopping needs of prime spenders become more geared toward home and children's fashion products. These conditions are what retailers such as Zara pursue by offering not only intergenera-tional product line diversifications but also home products and décor options.

An example of the strength of this dynamic comes from Ross and Harradine (2011), who surveyed 150 business students on their propen-sity to change brand allegiances when offered cheaper clothing substi-tutes. According to the authors, the substitute options had the same, if not better, product features as the branded garments. Surveying busi-ness students is an excellent approach to evaluating the strength of brand allegiance, because, as noted by research covered in Chapter 2, strong brand allegiance is associated with a lower degree of rationality. Business students are expected to have a relatively more rational nature than the average consumer. They are also expected to show a higher propensity toward monetary value. A third expectation is that business students pos-sess higher awareness of marketing manipulation. After all, the market-ing material they are studying overwhelmingly discusses the concept and problems of market manipulation. Therefore, one can expect that in an experimental study, the combination of knowledge of possible marketing

manipulation with the higher degree of information about the subject*
should result in the students being less susceptible to brand allegiance in
the presence of viable and valuable substitutes. Brand allegiance is a direct
outcome of successful brand manipulation. The results of the study indicate that even though the students noted a positive experience with the
substitute value brands, they indicated that this experience would not be
enough to change their purchasing habits. The implication of this finding
is that the importance of a label is much stronger than direct economic
gain. Even when significant satisfaction of taste and preference is achieved
by buying the cheaper substitute, even the most rational and budget savvy
of modern-day shoppers value the utility of an image brand more than the
combined utility of gained value and efficiency.

This fact is a strong testament that marketing manipulation works.
There is evidence that it starts early in life. An entire culture of consumption is based on the successful deployment of marketing manipulation,
and that culture is the foundation of our modern-day understanding of
economic prosperity. The more marketing stimulates us, the more purchases we make. The more purchases we make, the more demand we create. The more demand we create, the more production there is as a result of
that demand. The result is an increasing marginal growth of consumption,
which is the foundation of marginal increase in wealth creation. Therefore,
we define prosperity through marginal consumption. The question is: are
we paying the fair marginal price of increasing our apparel consumption?

UTILITY, ETHICS, AND THE QUEST FOR SUSTAINABILITY IN FASHION COMMERCE: NOT WHILE PRICES ARE FALLING

The quest to build intergenerational brand allegiance has implications
for the questions raised by analysts of ethical marketing and the promotion of sustainable fashion products. The research on brand promotion
to younger and younger consumers has raised questions about the ethical components of such a platform. Bergadaà (2007) states that marketers employ covert approaches in advertising that rely on young people's
lack of analytical understanding. A tactic discussed is the use of social

---

* Which leads to higher levels of rationality.

networking sites, through which an illusion is created that there is less direct advertising and more peer product evaluation input. The end result, however, is an increased propensity to buy the products that are the subjects of social media discussion.

Cassidy and van Schijndel (2011) call this tactic a "stealth approach" of advertising. Grant (2004) explains that marketing executives hold the view that young people are neither involved nor sophisticated consumers of advertising. They approach youth marketing weary of a lack of engagement and low literacy levels. In relation, while explaining the alleged low level of teenage analytical understanding of advertising content, Bergadaà (2007) argues that it is due to young people's lack of awareness of product features. Therefore, they are more susceptible to marketing manipulation. In summation, Bakewell and Mitchell (2003) offer an overview of the literature on youth consumption and provide evidence that the amalgam of all previously discussed factors has resulted in a reality in which Generation Y consumers have been socialized into habitual consumption earlier than previous generations.

With respect to fashion, Pentecost and Andrews (2010) find that Generation Y consumers have higher-priority attitudes toward impulse buying compared with older cohorts. Consequently, they have higher purchase frequency. Both these outcomes are dependent on the fact that Generation Y has high "fashion fanship" (Pentecost and Andrews, 2010, p. 26). Lin and Chen (2012) examine factors that contribute toward higher priorities on impulse purchasing, and find that adolescents are particularly sensitive to interpersonal communication. This is the case because today they grow up in an environment of communication that is defined by social media information exchange. Lin and Chen's results reveal that the greater the frequency of interpersonal communication, the greater the propensity to engage in impulse purchasing. Among the factors that drive that propensity is fear of immediate negative evaluation. This fear is connected to impulse purchasing in two ways. It shows that: (1) social media exposure and communication increase impulse purchasing and (2) peer pressure increases impulse purchasing not only in store environments, but also in online shopping.

The implications are sobering: that insecure teenagers who prefer impersonal communication, or for other reasons tend to engage in it more than others, are more impulsive. As already noted by the volumes of research connecting feelings of social inadequacy to clothes and brand allegiances, for these shoppers, the priorities of purchase go far beyond the regular definitions of utility. This behavior is one of their few control mechanisms, because such consumers use the purchases as building blocks of their own self-worth. Insecurity in consumption behavior is linked to societal unhappiness that has developed into neuroticism (Johnson and Attmann, 2009). Lee et al. (2007) define the targeting of such psychological needs as *neuromarketing* and posit that it merits its own specific field of study because its incidence of deployment is growing. Neuromarketing relies on building trust. That trust provides a support mechanism that helps offset insecurity.

Neurotic customers are the least likely to change their consumption behavior with respect to downscaling or substitution. This is a very important point to remember, because not only does it have ethical implications, with respect to the exploitation of their feelings of inadequacy, but also because it means that they have the least elastic consumption functions. These consumption functions are not only inelastic, but prone to continuous augmentation. As the shopping behavior of neurotic consumers becomes increasingly inelastic, so does their ability to increase marginal levels of utility. In other words, they shop often, and it is unlikely that they can increase the number of shopping trips, because they shop daily both in stores and online. But what they can increase is the number of purchases. Therefore, to make themselves feel better, neurotic consumers buy more. Johnson and Attmann (2009) explain that neurotic needs are *impulsively* chased and *religiously* met by consumers, because a lack of doing so leads to increasing levels of negative feelings.

It is because of such discussion in the fast-fashion retail discipline that ethical questions are raised that probe into the responsibility of marketers. Is it to increase sales at all costs, or is it to teach responsible consumption? The legacy of fashion marketing examined in Chapter 1 indicates that the industry teaches consumption patterns. The literature on demographic consumption patterns of modern fashion buyers examined in this chapter indicates that they have learned how to shop. The outcome is a low regard for responsibility and high regard for giving in to temptation. It is this succumbing to marketing stimulation that builds retail profitability.

Based on these findings, it can be concluded that learning of consumption patterns occurs through the combined input of advertisers and social environments. Bakewell and Mitchell (2003) clearly state that Generation Y and those following are brought up in a culture where shopping is not merely an act of purchase, but an act of entertainment and social excitement. Chen-Yu and Seock (2002) examine this social format by studying the clothing shopping habits of ninth to twelfth graders and find a degree of gender convergence. It is important to note their finding, because traditionally female customers, or in this case girls, had been the focus of fashion marketing.

Male fashion consumption had not been a main driver of industry commerce. The promotion of men's fashion and related products was vested with the older male customer—working men who dressed for the office. However, recent men's fashion research notes a heightened propensity for men to be fashion forward, and this research links the phenomenon to building the shopping habits of boys. Workman and Cho (2012) examine the proliferation of the male "metrosexual" phenomenon—a fairly new cultural occurrence in the constantly blurring lines of gender identity and behavior. Metrosexuals are young men who exhibit high regard for their appearance and spend significant amounts on clothes, cosmetics, and accessories. Workman and Cho (2012) find strong similarities between the buying habits of young men and women. Their results reveal that while minor differences exist between males and females when it comes to convenience, recreational shopping and fashion-consciousness (males tended to go for convenience above all), there are no behavioral differences when it comes to impulsive, quality, brand or price shopping orientations. This is an outcome of the dynamic Chen-Yu and Seock (2002) observation, 10 years prior to the 2012 study of Workman and Cho, that boys were equally socialized in clothes-shopping habits.

The amount of money modern boys spend on clothing is similar to the amount spent by girls, and they are equally stimulated by conformity, sexual attraction, and recognition motivations, or—in plain language—a desire to be noticed. These facts mean that boys today are a lot vainer than their fathers and grandfathers, and pay much more attention to fashion than previous generations of males. It is all a product of successful advertising. In the academic field, this phenomenon is covered by the literature on gender convergence in fashion purchases.

This body of research also makes important points with implications for marketing models. In spite of the general convergence, boys and girls differ in how they process fashion input from their peers. The difference lies in how they value the importance of social cohorts. Female teenagers value information from friends much more highly than male teenagers (Workman and Cho, 2012). This phenomenon is referred to as *peer-to-peer* marketing. Peer pressure is accepted as a defining factor in youth clothing purchases (Elliott and Leonard, 2004). In this context, peer-to-peer marketing exploits the social conformity tendencies of youth-group dynamics. Peer-to-peer market analysis explains how social gradation of such concepts in brand perception as "cool" are created and diffused among young shoppers.

No amount of direct or traditional marketing can come close to the affirmation a young person receives from his or her social circle. Therefore, for industrial psychology purposes, it is imperative to understand that the impact of dressing for a young person is subject to immediate audience approval. Decisions of personal style are, therefore, anything but personal. They are carefully dictated and changed through peer input. For that reason, peer-to-peer marketing, or—as it has traditionally been called—word-of-mouth marketing, becomes very important. In addition, today its magnitude is amplified by the degree of information channel diversification. The high diversity of information-exchange options among young demographics makes peer-to-peer marketing increase in scale and scope (Williams and Page, 2011). This diversity grows constantly through the modern advent of new technological platforms.

In this growth of opportunities for information exchange, a surprising claim comes from the field of fashion merchandising that studies the promotion of eco-fashion. Apparel consumers are not aware that their purchases create significant ecological damage. Morgan and Birtwistle (2009) find that consumers lack understanding of how their purchasing behavior impacts the environment. Bhaduri and Ha-Brookshire (2011) argue that this is because of transparency issues that range all the way through the production chain of clothes, where each link has become adept at hiding or not communicating uncomfortable truths about its ecological impact. For those reasons, although aware of issues of textile waste and growing fashion consumption, consumers do not see themselves as guilty of promulgating the problems.

Bhaduri and Ha-Brookshire (2011) note a sobering lack of connection in customers' minds that their behavior contributes to the textile waste issue. This lack of connection is understandable. It is because of two main reasons. One is that fashion retailers do not incorporate that information in their promotional platforms. The other is that the main message of those platforms is that, through spending, one shows care and support for the industry. The previously examined evidence tracking a consumer culture of learned purchasing habits has suggested that customers consider their spending on fashion as an act of reinforcement. Those consumers who love fashion feel that they are doing great things for the industry by fueling it through their dollars. The feeling is that the more they spend on clothes, the better. But because clothes today are cheap (in comparative terms), the meaning of "spending more" becomes defined by "buying more clothes and doing so more often." Fashion consumers and retailers do not see a problem with this behavior. These same consumers do not think of the pollution in clothes production, because they do not see clothes as polluting goods. If aware, retailers certainly are not going to readily reveal that fact.

Hill and Lee (2012) explain that this outcome is due to fashion marketing messages that are confusing. An additional component the authors discuss is the fact that sustainability and environmental impact are in themselves nebulous in interpretation. Hill and Lee (2012) explain that in 1987, the World Commission on Environment and Development (WCED) outlined in its mission a definition of sustainability in industrial production practices, which states that sustainability in production means addressing social, environmental, and economic concerns with the goal of preservation. This broad definition offers no concrete guidelines, and so companies interpret and implement its directives in many different ways. Therefore, communication of sustainability efforts to consumers is often ineffective (Du et al., 2010; Hill and Lee, 2012; Presas, 2001; Soron, 2010). Hill and Lee (2012, p. 477) state that in those fashion marketing campaigns that choose to incorporate sustainability messages,* these messages provide "information insufficient for the consumers' understanding of sustainability, leaving them with little knowledge of the effects of a company's sustainable actions." This is not just the reality of fashion advertising, but of environmental advertising in general. Brockington (2008) explains that this is because most environmental messages are not grounded in specific everyday realities of consumption.

The outcome is a lack of connection between environmental degradation and the frequent purchases of clothes. From the examined literature on the promotion of eco-fashion, it can be concluded that such promotion faces significant challenges in attracting consumers. The main reason is that the general public does not consider clothes to be polluting goods. This is the case in fashion sustainability promotion, but it is not the case with other comparably cheap, easily disposable goods. The general public today is perfectly aware that plastic bags, Styrofoam containers, and aerosol products cause serious environmental damage,

---

* For very few campaigns do so.

and the consumption of such products (although not ceasing) reflects a stewardship mentality, with high incidence of recycling and preference for "green products"—products that advertise environmentally friendly features (Lee, 2008a). A whole discipline of research has developed around green advertising, and much of it has stressed opportunities to target the teen market (Hartmann and Ibanez, 2006; Lee, 2008a, 2009; Peattie and Crane, 2005; Yam-Tang and Chan, 1998).*

Today, the largest and fastest-growing rates of environmental activism are among teenage and young adult demographics. These are the defining market demographics in fashion economics. These young people exercise their social consciousness values in consumption patterns across products. However, when it comes to fashion, there is little evidence of their commitment to environmental problems. This fact is surprising, because so much evidence exists that teenagers are the audience that embraces messages of the need for environmental stewardship with most alacrity.

As early as the onset of the 2000s, McCaughey and Ayers (2003) find that teenager social networks employed the largest incidence of environmental activism topics in relation to overall content. The authors explain that among the reasons is the fact that Generation Y has been raised with social activism. More recently, Williams and Page (2011) explain that Generation Y consumers are uniquely aware of many global issues, and this fact is mainly due to their high technological aptitude. Members of this generation also spend in accordance with such awareness and exhibit a propensity to support "green" products. Another important observation, which may be the defining link to understanding the paradoxical lack of connection between the environmental consciousness of today's youth and their unsustainable apparel-shopping habits, comes from Anderson (2011). The author explains that when it comes to environmental activism, the spokespeople have changed. At the outset of the promotion of environmental consciousness, scientists were the main spokespeople. Today, celebrities are. As already noted, in fashion the celebrity model is the main promotional tool, so to speak. The issue is that celebrities promote the sale of clothes and not their conservation. No fashion campaign employing a celebrity model includes messages of responsible apparel consumption.

Delving further into the role of celebrities in disseminating environmental consciousness, Anderson (2013) explains that institutional pressures

---

* These are some of the most-cited works in a very large body of research.

have made journalists—the traditional information disseminators of environmental information—more reliant on "prepackaged" information from the public relations industry. Therefore, celebrities have become increasingly important because of public relations promotion. They can bring an issue into the news spotlight because they generate newsworthiness with who they are. The paparazzi culture and increasing amounts of interest in the everyday activities of celebrities have, by definition, made celebrities a ready news event, to be used as needed. By making a small gesture, a celebrity can generate a news event much more effectively than a scientist or a politician, for example.

As early as 2003, Moeran noted that increasing amounts of celebrity cross-promotional activities had ecological overtones. Cross-promotional activities are defined as giving commodities a unified personality across media, into one integrated brand staple (Moeran, 2003). In the case of apparel advertising, Winge (2008) offers a thorough chronology of celebrity-championed modern "eco-fashion" promotion. The author concludes that even with emphatic celebrity support, eco-fashion options are only sellable if their esthetics are no different from the non-eco-fashion options. The point of the article is to show that modern eco-fashions are a far cry from the low-quality hippie garments of the 1960s. Winge (2008) takes an enthusiastic tone over the high quality and variety of clothing options that can be classified as "eco-fashions." Unfortunately, the article is nebulous in explaining what the concrete features of an eco-fashion-integrated production are. The classification suggests that the employment of organic cotton and environmentally "friendly" dyes is enough for a garment to be considered as a piece of "eco-fashion."

However, as much as we would like to embrace the concept, the evidence presented in this book suggests that there is no such thing as eco-fashion. There is organically grown cotton, but not organic cotton fabric. This is the case because of the "spinning" of the organic fibers into yarn, which is a process based on the heavy use of industrial bleach, peroxide, and acid—decidedly inorganic chemicals. Even the most environmentally friendly dyes employing the highest concentration of natural pigment are not really eco-friendly. For one thing, they are manufactured very inorganically—the pigment is extracted from its natural sources through intense production processes in refining that create toxic waste. For another, these "eco" dyes are applied to fabric through environmentally degrading methods that use 200 times the amount of water per one cubic unit of fabric. Therefore, eco-fashion, with currently employed technological methods, is not really an environmentally safe consumer option. It may be just a slightly less polluting option.

The eco-fashion call may have some customer support, but in light of the growth of fast-fashion retail outlets and volumes of research that suggest that customers do not possess clear information on the ecological impact of their fashion options, a discrepancy in the message is evident. This discrepancy may lie with the diffusion of the message in what Moeran (2003) calls *the cross-promotional name economy*. "Name" refers to the integration of several brands in one cross-promotional schedule under the image of the one celebrity who is the spokesperson for those brands. Terasaki and Nagasawa (2012) explain the interplay in terms of promoting environmental consciousness. The authors state that through providing visibility on certain environmental issues, celebrities build their own and unique ecological images. In this way, celebrities create an "issue brand." In the end, it is that image that the consumer digests, as opposed to the individual causes or products the celebrity represents.

This fact is the link that explains the information void of environmental damage in fast-fashion production and its increasing consumption. It is celebrities who promote fast-fashion brands today. Their target audience is largely comprised of teenagers. These celebrities have already established their own personal "brands." As they build themselves as brands, celebrities increasingly engage in promoting the fact that they support social and environmental causes. Therefore, teenagers do not think to question their credibility. For example, supermodel Gisele Bundchen, megastar Madonna, and David Beckham have promoted H&M. All are known for their social activism on issues from saving the rain forest—as a native of Brazil, Gisele is an avid advocate of rain-forest conservation—to poverty alleviation, which is a cause Madonna supports. A few recent examples of how those specific celebrities use the media to build their images as environmentalists show how they choose to focus on certain information that portrays them in the most favorable light as social activists.

In a cover story for the 2015 March issue of British *Vogue*, Gisele Bundchen states that despite her incredible wealth, she is most comfortable barefoot playing in the grass with her children. She goes on to explain that she feels most connected to nature and would live in a tree house if it were feasible. A significant portion of the interview is dedicated to her love for nature and work on environmental and conservation issues (Kilcooley-O'Halloran, 2015). However, the story also reveals that the tree-house-loving "environmentalist" supermodel made more than $47 million in

2014. A significant portion of that money is income earned as a spokesperson for H&M. However, most of Gisele's wealth does not come from modeling but from her own fashion-line businesses. Gisele is herself a fast-fashion industrialist, owning a few clothing MNCs headquartered in her native Brazil that specialize in swimwear and activewear—decidedly not eco-fashion product lines. That part of Gisele's fast-fashion industrial involvement, however, is not discussed in the article. The main message is that the ultrarich* beauty deeply cares for the environment. She has an institutionalized track of environmental activism in the fact that she is a goodwill ambassador of the United Nations Environment Programme and also sits on the board of the Rainforest Alliance—a nonprofit organization dedicated to conserving biodiversity and sustainability.† With such a strong image of an environmentalist, nobody would suspect that any of the products Gisele endorses or, even worse, owns as an industrialist herself cause serious ecological damage.

Another example is the visibility of social justice issues Madonna's work in Malawi creates. This work is in community development, which includes addressing issues of clean water and environmentally safe food production. These social causes are ecological in nature. Therefore, the brand of her celebrity also becomes one of an environmentalist. She stresses this role with public appearances such as performing at the launch of the Live Earth concert in 2007. Live Earth is an entertainment series platform created by award-winning producer Kevin Wall and former US vice president Al Gore to increase environmental awareness through the power of entertainment. Kevin Wall was also the mastermind behind Live Aid, the famous 1984 multiartist concert for poverty alleviation, where Madonna also performed. The megaproducer is reviving the Live Earth concert series in 2015 with Live Earth Road to Paris—a series of concerts in several major cities around the world, to be held in the summer of 2015, that aims to build momentum for the upcoming December 2015 UN Climate Change Summit to be held in the French capital. Madonna is taking part in these performances and is recording a special song for the event in collaboration with hip-hop singer/songwriter Pharrell Williams (Pantsios, 2015).

---

* It is reported by celebrity.com that Gisele is the richest supermodel of all time, with a net worth upward of $320 million, due to her successful fashion entrepreneurship in successfully launching a few fast-fashion labels, mostly sold in Latin America.
† As reported by *Forbes* Magazine on June 10, 2014. See http://www.forbes.com/profile/gisele-bundchen/.

The core information that consumers digest when they see celebrity fashion models in such social roles is that as spokespersons, celebrities such as Madonna and Gisele represent products that are environmentally safe. Therefore, teenagers flock to H&M, never having a reason to suspect that the cheap fashions they purchase there create significant ecological damage. It is a clever promotional tactic that magnifies the message that H&M's most famous celebrity models—Madonna and Gisele—have strong personal "issue brands" as environmental activists. Therefore, the company benefits from the transference of the images of its celebrity spokespeople "issue brands" into its own image. The final outcome is that H&M appears to be a "green" company.

Is it a coincidence that the most environmentally visible stars of global pop culture are recruited to promote the most pollution-generating apparel product lines? Or is it a cleverly calculated choice made by H&M promotion executives to recruit celebrity models with high environmental cache? Would Gisele and Madonna put their faces on a line of disposable plastic bags or Styrofoam cups? It is highly unlikely. Perhaps this fact of promotional message distortion is behind the paradox that as sales of fast-fashion brands increase, so does the incidence of eco-fashion promotion. The problem is that these two concepts are seldom linked by analysis.

Activism continues to grow around the concept of socially conscious fashion production. Cervellon (2012) explains that not-for-profit (NFP) campaign companies engage on their own in investigating environmentally damaging practices. Upon discovery, they spearhead emotional campaigns of strong messages that rely on shock value. Cervellon (2012) offers examples of such shocking language as "Donna Karen Bunny Butcher" and "Bloody Burberry." However, the impact of such campaigns is limited, because they are polarizing. Their sources are easily discounted as ideological zealots. As explained before, there are political factors that influence eco-message absorption in the general public. Those customers who define themselves as politically conservative would not respond to eco-messaging. Evidence of the limitations of socially conscious and eco-friendly promotion in fashion suggests that it is unclear whether these messages impact customer decision-making. There is actually empirical evidence to the contrary.

The previously discussed results of Ellis et al. (2012) suggest that customers who did not earn their fashion-purchase-allotted income were most willing to pay ecological premiums. These are nonwage earners, and as previously explained, they have relatively lower elasticity of consumption.

Counterintuitively, those customers who earned their money and even exhibited the sociodemographic signs of being in nonconstricted financial situations were not willing to pay (Ellis et al., 2012). The implication is that older and more aware customers are not convinced by the "organic" rhetoric. Younger customers are, and that is a hopeful sign for future development of fashion product lines.

Even so, the conclusive summation of the empirical evidence presented in the above-examined literature suggests that fashion consumers are not swayed by passionate attacks on brands. Brand allegiance is, indeed, a strong phenomenon that has withstood criticism, because it is based on cultural platforms that have been gradually augmented for generations. These platforms include the promotion of habitual consumption and the importance of convenience, both in access to products and in learning about them.

Naomi Klein most famously discusses the negative side effects of strong brand allegiance in the famous 1999 book *No Logo*. Klein tracks brand promotion and manipulation for general consumer products, but spends considerable time on clothes, because she puts brand manipulation in the context of the 1980s and 1990s mall culture, where clothing purchases defined branded retail. Those were the decades before fast fashion, and during that time, teenagers bought relatively more expensive clothes. Their brand allegiance back then contributed to the high profits of fashion houses and fueled the influence of the few megafamous fashion labels on the entire industry. Today, fashion retail remains a branded industry; however, there are new competitors in the market. Their ability to compete on price, while skillfully developing their own brand legacies, has changed the industry, but without diminishing the role of brands. From the extensive research examined in this book, it is apparent that there is little evidence of brand desertion. Hyllegard et al. (2012) offer reasons why this is so, arguing that it is clarity of information as perceived and digested by consumers during the shopping experience. Consumers pay attention to information on hangtags and actual clothing labels. The authors argue that if the information is not there, consumers do not exhibit signs of favoring socially consciously produced clothes. The main information on labels is the brand name.

8

Implications and Conclusions

THE ECONOMIC REASONS FOR IGNORING SOCIAL COSTS

The implication from the conclusion of Hyllegard et al. (2012) that information on the eco-features of clothes needs to be on hangtags and actual clothes labels for consumers to respond raises feasibility concerns. A main question is whether that is a wise recommendation, given the volumes of research on integrated marketing that examine the multitude of information streams deployed in fashion advertising. It may be true that at the point of sale the information on the labels and hangtags has the greatest impact on choosing which garment to purchase; however, the decision-making process that gets customers to the point of purchase is influenced by many and varied sources of information that all discuss apparel attributes. It is in these sources that the discussion of ecological damage is absent.

The legacy of integrated marketing research (as tracked in Chapters 2, 6, and 7) has built a general platform of understanding that apparel promotion has evolved from unidirectional to user generated (Kulmala et al., 2013). Unidirectional promotion is based on trend dictation from fashion professionals to the general public. User-generated (or, it may be better put, user-aided) promotion relies on information from the general public on the features of apparel products. Neither presents the truth that the production of mass-market apparel is highly ecologically damaging.

There are two reasons that have to do with the direction of information in apparel promotion. One is that the unidirectional legacy of branded retail promotion lies in a lack of incentive to communicate that truth to the consumer. The other reason is that consumers do not understand how clothes are produced, which is not unusual. Therefore, they do not talk about it, so the topic is not reflected in user-generated fashion content.

Consumers do not understand how most goods are produced, because modern-day technologies employ highly sophisticated methods. However, consumers tend to have a general idea of what goods are relatively more pollution intensive. This is the case because manufacturers of pollution-lessening technologies promote the features of their innovations to the consumers in search of building competitive advantage.

In apparel production, that dynamic is not present. The main reason is that there is no real technological innovation sector in textile manufacturing that is committed to the development of pollution-mitigating technologies. The innovation in textile production is focused on producing high-quality, high-choice, low-cost fabrics. Known as *poly blends*, these fabrics define the ability of modern garment producers to increase retail sales by offering higher levels of choice and lower prices. Because sales are growing at rates much higher than historical averages, the industry does not face incentives to change production platforms.

Sales are growing because customers are responding to the allure of low prices and high choice. Thanks to fast-fashion apparel options, today, customers can create a "million-dollar" look for the price of 10 dollars. One cannot fault customers for wanting to look like a million dollars in a culture that so avidly tells them they need to. Ten years ago, if one shopped at Walmart or Target, one looked like it. Today, one doesn't have to. Even with the right amount of public information, education, and promotion of purchasing truly sustainable products, a reasonable analyst would not expect low-income consumers, in particular, to give up that kind of utility. Today, one can be on welfare and not look like it. The cultural utility of such a reality would trump all pressures to shop responsibly and pay the fair price for environmentally damaging products.

For these reasons, information on the ecological impact of garment manufacturing is not communicated to the consumers from any production-level sources. Such information comes from nonindustry sources. It is, however, disjointed, confusing, inconclusive, and ambiguous, because it skirts around the main factor of environmental damage from the apparel industry. That factor is the promotion of overconsumption.

As early as the late 1990s, attention was raised to the issue of overconsumption. Buyer behavior such as "ethical consumption" was noted (Shaw and Newholm, 2002). Ethical consumption is downscaling, or contracting, individual consumption as a counteracting response to consumer-oriented culture. At its core, the concept is alien to modern-day apparel retailing. Today's apparel models are all based on stimulating impulse purchasing. The literature sums up the admonition that in order to be successful in business today, retailers must not only stimulate customers to buy, but overstimulate them.

Overstimulation relies on surprise. At its core, it is vested in the assumption that even when customers are in stores and have a certain level of planned purchasing in mind, overstimulation will entice them to make additional and unplanned purchases. The literatures on brand proliferation, service management (with focus on apparel sales), fashion merchandising, and retail operations management all conclude that the successful model for apparel retail growth lies in increasing sales through providing exciting in-store (or online) experiences for the purpose of enticing. As a result, the shopping experience for clothes is managed as an entertainment venue. Its main goal is to augment individual consumption, not to contract it. Therefore, "ethical consumption" in apparel is skillfully morphed not toward contraction of the amount of goods sold, but toward promoting the consumption of apparel products that are differentiated via their "eco-fashion" attributes. The sad reality is that there is no such thing. *And this fact is what industry insiders skillfully hide in their advertising communication strategies.*

It is understandable that industry professionals do not want to hurt their bottom line. They are just following the accepted model of economic prosperity that drives all industrial sectors: that is, the model of sales growth. However, growth is subject to costs, and there are two types of costs—direct costs and social costs. In the apparel industry, the social costs are those of environmental damage, which, as explained in this book, occur at all links of the production chain.

Business institutions are expected to be vested in the minimizing of direct costs. Social costs, to them, should be fixed costs, that is, the costs of doing business. Fixed costs bear their name because it is assumed that they are not subject to mitigation or avoidance. In other words, businesses cannot choose not to pay them. Fixed costs are those of purchasing necessary equipment, investing in infrastructure, and paying legal or other fees that are necessary to obtain industrial permits. Pollution mitigation technology and costs associated with environmental regulatory compliance, therefore, should be fixed in nature. However, they can only be fixed in efficient markets, or, as is understood by economic theory, markets that are relatively free and unencumbered by government intervention.

As noted in Chapters 3, 4, 5, and 6, the international market of apparel production is anything but free from government intervention. It is highly subsidized at the production end and highly monopolistic at the retail end—two conditions of market inefficiency.

In inefficient market structures, social costs can be mitigated through the power of organization in two ways. One is that those producers that create social costs can ask the government to bear the costs through subsidizing them. The other way is through winning agreement from the government to absorb the negative impact of social costs by tolerating environmental damage. It is a trade-off that exists in all industries, because all industries create pollution.

As the social cost concept grew and penetrated economic analysis, most notably with the proliferation of environmental legislation post 1960, social cost mitigation was deemed to fall within the scope of producer and government interaction. As environmental laws became part of the world's legal system, their implementation and adjudication remained within the scope of governments. The key issue here is the plural nature of governance. National sovereignty dictates that each nation is free to create, implement, and regulate its own legal system. Therefore, in each nation a different government is free to decide how to deal with social costs. The problem is that in one globalized market, this reality creates incentives for organized international commercial interest groups to lobby governments for preferential platforms that provide effective social cost absorption. In this way, social costs morph from fixed to variable costs for businesses.

The evidence examined in Chapters 3, 4, and 5 points to the fact that this dynamic is very much at play in the apparel industry. Nations that rely on apparel production strive to provide the most attractive business environments for international apparel conglomerates. Among these attractive features are lax environmental standards that allow apparel MNCs not to pay for costly pollution-mitigating technologies. As a result, the most environmentally taxing links of the clothing production chain occur in the very poor nations of the developing world. The benefits of this social cost mitigation, however, accrue to the customers in all markets in the form of low retail prices.

The literature on unequal ecological exchange explains that as nations grow richer, their preference for public goods, which include a (relatively) clean environment, grows (Jorgenson, 2006, 2007, 2009; Presas, 2001; Rice, 2007; Shandra et al., 2008; Wagner and Timmins, 2009). Public goods are those quality-of-life improvements that governments provide, such as a clean and safe environment, national defense, safety, security, and education programs. In developed nations, regulatory structures stringently regulate industrial activity, and the richer the society, the higher its provision of public goods with environmental impacts (Bateman and Willis, 2001; Kahneman and Knetsch, 1992; Sagoff, 1998). However, in developing nations, where regulatory institutions, like all institutions, are by definition developing, volumes of literature find that environmental public good provision is negligible, and that fact attracts foreign investors. On the whole, these investors are firms from nations with stringent environmental standards that are looking to evade them in order to save on direct production costs (Grimes and Kentor, 2003; Jeppesen and Hansen, 2004; Jorgenson, 2007; Lee, 2009; von Moltke and Kuik, 1998). Falling within the scope of the literature on unequal ecological exchange, this body of work develops the pollution haven hypothesis. The pollution haven hypothesis states that FDI is the direct tool through which rich nations pass on their social costs to poor nations. Based on the data examined in Chapter 5, supporting evidence in textile-intensive economies of the pollution haven hypothesis is best illustrated by Figure 8.1. Values of water pollution from textile manufacturing are plotted from three types of nations that best define the economic features of countries significantly involved in the apparel trade. They are: (a) major textile exporters close to a major market—East European nations that are the main suppliers

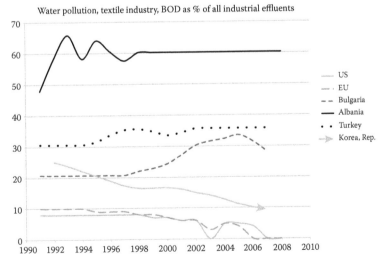

FIGURE 8.1
BOD pollution as percentage of all industrial water pollution for selected nations that are leaders in global apparel commerce.

of both textiles and apparel for the Western market, (b) the Western market itself—the United States and the European Union (EU), and (c) a developing nation that has industrialized to the point of becoming a net foreign investor—South Korea.

Among the main arguments of the theory of unequal ecological exchange that developed the pollution haven hypothesis is that industrialization growth causes increasing preference for higher-quality public goods. However, instead of pricing the provision of those public goods internally, nations tend to outsource the dirty production of industrial goods to poorer countries. This process of externalizing, to put it in economic terms, social costs is a result of retail-level competition on direct costs. This dynamic is apparent in the case of South Korea, as an example of a nation that has industrialized rapidly in the recent past. Figure 8.1 shows how the water pollution from textile processing has decreased significantly. As explained in Chapters 4 and 5, South Korea developed its textile sector to such an extent that it became a leading source of global textile FDI, which it directed mostly to African nations. Much of the evidence examined in those chapters explains that the main goal of the investment was to augment MFA quotas. It would be naïve to think that quotas were the main driver for this site-selection process. Externalizing the highly polluting processes was also at play, and this continues to be the reality in all developed markets. The data from the United States and the EU offer additional support for this line of argument. In the largest retail markets, continuously growing because of fast-fashion sales, water pollution from textile manufacturing is decreasing. Meanwhile, in their major trading partners, which are increasing textile export volumes most significantly, water pollution levels are growing. The data from Albania, Turkey, and Bulgaria offer support for that point.

It is apparent from the data in Figure 8.1, and from the literature across disciplines that discusses environmental consequences of globalization, that in fashion economics, environmental sourcing is very much at play. It is the established industrial platform, without an alternative. The competition on price is so strong that it has created a system of intricate international trade based on cost-minimizing site-selection policies as the drivers of competitive advantage.

In international trade, competitive advantage refers to national economic features that define countries' strength in producing certain goods. Knowledge-intensive economies, that is, developed nations, tend to specialize in knowledge-intensive goods, while labor-intensive economies, that is, developing nations, specialize in the production of goods, where the value is derived from the abundance of cheap labor or land. On the whole, modern international trade is in exports of durable goods, such as machinery, industrial equipment, robotics, airplanes, and arms, that developed nations sell to developing nations that do not have the technologies to produce such goods. Developing nations, in turn, mostly export commodities and agricultural products. At least, that is the theory of efficient international trade based on factor-endowment assumptions. These assumptions state that national competitive advantage develops around natural endowments that influence factors of production. For example, nations with abundance of land would export agrarian products. Nations with abundance of capital would export knowledge-intensive goods. However, in the international trade of clothes, none of those assumptions holds.

In the most internationalized of industries, the richest nation in the world, the United States, exports cotton—an agrarian commodity. It also exports its retail business models. It could be exporting the equipment needed in modern textile processing plants, but this dynamic is not reflected in any research or data on fashion and apparel component exports. As well as exporting cotton, America has also gradually exported most of its pollution-intensive textile manufacturing processes. And so have other developed nations, as the literature on textile FDI examined in Chapters 4 and 5 indicates. This process of exporting social costs is based not only on direct cost-minimizing incentives at the firm level, but also on government action that aims to help developing nations improve their economies. This action comes in the form of industrial recruitment policies. These are policies for attracting businesses. They are supported by international governance structures whereby multilateral poverty-alleviating organizations, namely, the World Bank and the IMF, provide loans and other incentives to the governments of developing nations in the hopes of increasing industrialization and, therefore, alleviating poverty. This well-intended policy direction has negative unintended consequences, or perverse incentives.

SOCIAL COSTS AND PERVERSE INCENTIVES IN FASHION ECONOMICS

As developing nations increase their industrialization impetus, they also diversify from dependence on agrarian exports toward labor-intensive manufacturing exports. Such exports are mostly inexpensive and easy-to-produce fast-moving consumer goods, chief among which, for decades, has been clothing. This diversification is executed with foreign funds, because local industrialists do not have the technological know-how to create a competitive home-grown alternative to any established international industry. Furthermore, they do not have the global market know-how to allow them access to viable foreign markets, which are necessary for the successful export of any product. After all, exporting can only be profitable if there is an international market for the exported goods. For these reasons, as developing nations climb the industrialization ladder, they increasingly rely on FDI for funding. The process is well studied by the literature on structural adjustment programs. As explained in Chapters 4 and 5, structural adjustment programs are loan programs administered mostly through the World Bank's International Bank on Reconstruction and Development and the IMF. They provide not only funds, but also economic training to governments in the developing world on how to attract foreign investors. Sadly, among the main conclusions in that literature is that structural adjustment programs create incentives for environmental exploitation.

The data on FDI flows in nations heavily reliant on both structural adjustment loan programs and textile exports presented in Chapter 5 offer empirical evidence of this reality. The most direct damage is in water pollution levels, which in certain nations like China and Bangladesh impact entire ecosystems, creating serious public health problems. Still, with all the evidence presented in Chapter 5 on that reality, developing nations continue to vie for FDI, and their ability to attract additional investment and, as a result, increase their exports is the measure of success in structural adjustment program evaluation. According to the structural adjustment program literature, pollution problems are not included in such evaluations. The reason is Western ethnocentric self-consciousness and fear of infringing on national sovereignty issues.

It is left up to individual nations to legislate their own internal laws, and environmental standards fall within the scope of these laws. The problem is that policy legislation is a function of incentives. If the economic incentives of a nation include competing for investors who are looking to minimize direct production costs, then all factors that mitigate direct costs become subject to decrease or elimination. As the literatures studying the pollution haven hypothesis and the theories of unequal ecological exchange note, among the main factors that offer MNCs the ability to lower production costs is not paying for expensive environmental cleanup. In certain industries, these costs are defining pillars of average variable costs—the costs incurred in production expansion—and their importance is a function of price. The lower the price point options of the final goods are, in terms of retail pricing, the higher the average variable costs in pollution mitigation become. This inverse relationship exists because the low retail-level prices do not allow producers to internalize the negative externalities of pollution. The way to internalize the costs of such externalities is to pay the fair price for environmental cleanup and factor in the price increase in the retail structure to the end consumer. Sadly, in apparel production, it appears that this dynamic does not materialize. Retail price competition is so strong that it does not offer internalization options for modern garment producers with respect to price markups that consumers are willing to pay. The evidence from the literature on eco-fashion retailing suggests that those retailers that experiment with price premiums through eco-labeling face major challenges in attracting consumers. Among the reasons examined for this lack of enthusiasm is a lack of customer awareness of the true social costs of the production of affordable clothing.

A major contributor to this lack of awareness is a legacy of regulatory policies in international fashion economics that provides cost-minimizing incentives to individual production links. The main type of such policies is the subsidizing of exports. A successful outcome of such a policy is inadvertently defined by the growth of exports. This is the case in developed and developing nations alike. In developing nations, as noted, exports are subsidized through structural adjustment program policies. In developed nations, however, the subsidizing platform is more intricate and twice repeated. Most notably, in the United States, the export of raw cotton is seen as a great success in agricultural efficiency. Therefore, as noted in Chapter 4, cotton exports are highly subsidized. But there is also an additional tier of subsidization in the form of the preferential rule-of-origin clauses of the US Import Code. Called the "product-sharing rule" of the import code, this subsidy comes in the form of preferential import tariff policies on textile products (either fabric or finished clothing) that are manufactured from American cotton. In the tightly guarded American retail market, such goods are given duty-free entry into the United States, which amounts to significant savings.

These two policies offer a double subsidy for the crop: at the beginning of the apparel production chain and at the end of the production chain.

The problem is that they incentivize the exportation of the pollution- and social-cost-intensive links in the apparel production chain. These are the weaving of fiber, the manufacturing and dyeing of textiles, and the potentially exploitative, cheap-labor-dependent assembly of clothes. In addition, they create the long supply chains discussed throughout this book that make the transportation of apparel among the worst carbon-intensive links in fashion commerce.

The perverse incentives from those policies go largely unnoticed because they are based on a concept in international trade that is accepted as a good thing. That concept is trade liberalization. Although the legacies of protectionist policies are the seeds of the current high level of internationalization of the production function of fashion goods, it is the impact of fairly recent trade liberalization policies that has eased international operations for apparel MNCs. However, as the evidence in this book shows, this agility of increased geographic choice enables fashion conglomerates to exploit institutional weaknesses in environmental regulation. Environmental regulation in the industry impacts mainly water pollution, and a successful institutional platform demands that apparel producers use costly mitigating technologies. In certain cases, as is the reality in the United States, it is not only content of liquid effluents that is regulated, but also volume. This fact goes largely unnoticed in the research literature on environmental sourcing.

Environmental sourcing is a concept that describes the process of choosing to locate production in areas with the least environmental regulation. The "sourcing" in the term refers to seeking out institutional loopholes. In water-intensive manufacturing processes, which include textile production, this process of looking for locations with lax environmental standards includes a capacity component. In America, water-intensive production is regulated mainly on volume. This is a legacy of the pulp and paper industries, which were the most polluting sectors during the time when the US EPA was established. Since then, the regulatory structures have evolved to include a level of discretion exercised by individual American states, where effluents are regulated on content—mostly BOD values, but also on volume (Wall and Hanmer, 1987). There are only so many gallons of water that American firms are allowed to use in a day. In Europe, on the other hand, water pollution is regulated mainly on content (Laing, 1991). The difference has to do with: (a) cost of regulation—as noted, it is very expensive for any government agency to conduct content level water toxicity tests; and (b) the decentralization structure of governance.

In America, individual states demand higher autonomy from federal regulation and use that autonomy as a competitive attribute. To a lesser degree than in developing nations, American states use their discretion in interpreting federal policy directives so that they vary from state to state. As a result, certain states have more stringent content regulatory structures, while certain states have more stringent volume regulatory structures. However, neither offers enough latitude to accommodate the capacity demands of modern-day textile production for mass-market retail. It is this capacity that defines the competitive advantage of textile producers in developing nations. They can take on large orders, use as many gallons of water a day as required, and not follow the demanding toxicity regulatory platforms implemented in the West to protect aquifers. As the literature on environmental economics that focuses on intercountry comparison of preference for public goods concludes, in developing nations, public health awareness drives the relatively high level of water-basin protection. That protection is very much valued, because industrial effluents travel through the groundwater system and pollute the potable water sources used in daily consumption as well as the food supply. Nobody wants a textile plant emitting thousands of gallons of heavy metals, bleach, and acid-laden effluents in their respective watersheds.

Even if producers followed all the regulatory platforms and engaged in environmentally responsible production of textiles, the hardest truth of all must be told. This is the fact that, as Kant (2012) explains, the toxic components discharged in modern-day textile processing cannot be removed through purification. They can only be mitigated slightly. Out of the 72 elements identified by industry standards as poisonous, only 34 can be treated through modern purification technologies. The treatment only allows the partial decrease of their levels, but not efficient removal. For these reasons, MNCs, mainly in textiles, have an incentive not to deal with stringent and interest group-driven environmental regulation in the Western world, but to set up operations abroad.

These incentives to internationalize operations stay under the favorable cover of policy rhetoric that exalts the benefits of free trade. It is assumed that free trade increases efficiency in production, which ultimately results in lower prices and higher choice for the consumer. These benefits are evident in modern-day fashion economics. Consumers today face unprecedentedly high choice in apparel goods and inconceivably low prices. The problem is that these benefits are based on perverse incentives to evade the internalization of the industry's high social costs.

TELLING THE HARD TRUTH: WHY EXTANT SOLUTION RECOMMENDATIONS ARE IGNORED

The apparel manufacturing supply chain dynamics reviewed in Chapters 4 and 5 reveal that mass-marketed garments are produced in ways that are energy inefficient and pollution intensive. The findings presented in those two chapters are not new. There is an extensive literature, which is cited copiously in this book, from several disciplines that discusses that unfortunate reality. Yet, little is changing, and it may be even that things are deteriorating. The reasons for this are as many as there are links in the industrial supply chain of fashion. They range from policy legacies, to lobbying and trade-distorting protection, to special and vested interests in the status quo, to the growth of a global pop culture that has changed both the nature and the lengths of fashion trends.

The proposed solutions to dealing with the rising pollution problems from the growth of fashion sales are just as numerous and varied as the causes of environmental damage. Daniel Sumner (2003, 2006, 2007) thinks the key is the abolition of all cotton subsidies in the Western world. Ten years after such calls, no subsidies have been reduced. Elizabeth Rosenthal (2007) thinks we need to wear more polyester, because it is the processing of cotton fibers that is most environmentally taxing. However, polyester by definition is a petrochemical product, and its manufacturing is no cleaner than that of cotton-based textiles. Rob Gray (2006) proposed the development of social and environmental accounting to help create sustainability value in a manufactured product. No such notable initiatives have gained ground in fast fashion. Feriggno et al. (2005), as well as Ferrigno and Lizaragga (2009), are among the best proponents of organic cotton, outlining its benefits for the poorest of nations, while Eyhorn (2007) argues exactly the opposite—supporting organic cotton production in underdeveloped nations creates more harm than good. But organic farming of cotton is only one link in the long supply chain of the cotton manufacturing process. It is part of farm-level costs, and as explained in Chapter 4, when it comes to final price levels, farm-level costs account for less than 6%.

Yet, the field studying the commerce of used clothing offers another solution. The notion is to promote the used clothing and repurposing sectors, because they can reduce the carbon footprint of new production. Farrant et al. (2010) offer empirical estimates and posit that the commercialization of one reused cotton T-shirt can reduce pollution directly related to global warming by 14%. The reuse of one pair of pants can reduce such ecological damage by as much as 45%. This is an actual policy recommendation that may have implementation merit because it targets the end consumers. The previously listed solutions ask industrial entities to change the way they do business. However, since business is so profitable as it is, these sound and meaningful recommendations are understandably ignored, for the following reasons:

1. The recommendation to abolish cotton subsidies is idealistic, at best. Cotton subsidies will never be abolished. Too much wealth is concentrated in the hands of a very few powerful and well-organized interest groups as a result of the subsidy system. Volumes of work in political economy examining every industry in every nation conclude that if interest groups have been successful in extracting subsidies from government, then those subsidies never disappear. This is the nature of rent-extracting and rent-protecting behavior, and it is contingent on wealth. The richer those subsidized are, the more successful they are at protecting their government-extracted rents because they can afford the high costs of lobbying.

   However, what could happen is the dismantling of the product-sharing rule system of the US import code in America and the preferential rules-of-origin classifications in the EU that provide cost-saving incentives in reimportation (as is the case in America). The likelihood of such a policy coming to fruition is also low, because it impacts retail-level profit margins. If preferential rules-of-origin policies ceased, retailers would have to raise prices, and that is not a popular notion in today's reality of competing on price.

2. The other recommendation to limit ecological damage is to support more synthetic-fiber-based fashions. However, it will be very hard to convince consumers to wear more polyester. The trend and style legacy of the past (remember bad 1970s fashions) has created a culture in which polyester is considered old fashioned and cheap, not in the good sense of offering high value, but in terms of quality and performance. Birnbaum (2008) explains that fast fashion depends on mostly cotton-based fabrics and some poly blends, because industry promoters cannot call any garment "fashionable" if it is made from polyester.

3. The most frequent and strongest recommendation is to support "eco-fashions." This recommendation is the most dangerous of all, because there is no such thing. Even discussing "organic fabric" is irresponsible and deceitful, because there is no production method that can create such a fabric. There is organic cotton, but not organic fabric. When a shopper culture of limited capacity for information absorption, as discussed by Hyllegard et al. (2012), tends to pay attention to eco-labeling above all (the notion that the information must be on the label or on a hangtag attached to the garment), putting the words "eco-fashion" on anything is akin to lying. It is like saying that light cigarettes do not cause lung cancer.

However, using unclear language such as "sustainable fashion" and "eco-labeling" is a convenient direction to take in addressing the ecological issues in the industry, because it offers a comforting message that does not jeopardize sales. It offers a platform for the *promotion* of sales. It is the sales of new product lines of apparel that are offered as a choice to the consumer. Those customers most concerned about the environment are expected to make a choice by purchasing "eco-fashions." Those customers not convinced, unaware, or mainly concerned with price would choose otherwise. The problem is that, according to the research examined in this book, those are the overwhelming majority of customers. This is the case because there is a clear division in the clothing market between customers who are concerned about quality and customers concerned with price (Hayes and Jones, 2006). Evidence suggests that the proportions are uneven, and there is a significant crossover or even exodus from quality toward price. The reason is that price is no longer a signal for quality.

These facts do not mean that industry analysts should give up on the quest to improve the current state of ecological damage resulting from the integrated commerce of fashion and related goods. However, they have a responsibility to be honest about the possibilities, because only honest analysis can lead to viable policy recommendations.

Policy recommendations also need to be feasible. Criticizing cotton subsidies, recommending synthetic products, or recommending support for nebulous eco-labeling initiatives is naïve and, as the results of the response from the industry to such admonitions in the past decade indicates, fruitless. It is fruitless because such policy recommendations do not provide clear incentive structures for producers.

In conclusion, based on the body of work started in the early 2000s that tracks the social movements for ethical or environmentally conscious fashion production, it is apparent that those movements disseminate information that is disjointed and based on unclear definitions. The much-used term "ethical fashion" means different things to different people, because there is no industry standard or definition for it (Joergens, 2006). Joergens (2006) focused her research on how the general public, as embodied by the demographic most likely to embrace the concept of ethical fashion, even understands the meaning of the term. The author built samples of European youth aged 20 to 26—as explained by the literature on youth social activism, the most proactive segment of world youth. European young adults are the most progressive, environmentally focused, and ideologically vested in ecological conservation because of an education system that has embraced and taught sustainability curricula since the 1980s (Sterling and Huckle, 2014). With that fact in mind, Joergens (2006) found a surprising lack of understanding of what eco-fashion is. The results from the multitude of surveys indicate that on the whole, the understanding of what ethical fashion production incorporates takes on elements from fair trade, social justice, corporate social responsibility, and environmental stewardship. But it relies on incomplete elements from each. In tracking the reasons for the nebulousness, Joergens (2006, p. 361) offers an attempt at defining ethical fashion as: "ethical fashion can be defined as fashionable clothes that incorporate fair trade principles with sweatshop-free labour conditions while not harming the environment or workers by using biodegradable and organic cotton." This definition does nothing but offer more ambiguous rhetoric, if one takes into account the overall information previously examined in this book.

It cannot be stressed enough that there is no such thing as not harming the environment when it comes to textile production. There is always going to be significant harm, because current technologies cannot mitigate the pollution from the farming of cotton and other natural fibers to their ginning, bleaching, mercerizing, spinning, and weaving, and eventually dyeing the final fabric. Even using the words "not harming the environment" is dangerous and socially irresponsible coming from industry scholars because it provides a fake sense of security for the end consumer. If an industry specialist indicates "no environmental harm" that means that there is such a state, and there *is not* such a state.

The second main issue with this definition, and others like it, is the mere concept of "organic cotton." The growth and harvesting of the plant only account for less than 6% of total garment production costs. The majority of environmental degradation happens after the cotton is harvested, because it must be chemically altered from its rough yellow-tinted state into marketable yarns priced on finery. The finer the yarn, and the better the quality of the final fabric, the more energy and water intense the bleaching, mercerizing, and spinning processes need to be.

Just turning the baled organic cotton into yarn creates more environmental water damage than the pollution associated with the pesticide and herbicide use in nonorganic cotton production. Furthermore, as water becomes a precious commodity itself, one should not forget that organic cotton is three times more water intensive than the commercial and GM varieties, offers lower yields, and can provide incentives to alter natural watersheds for irrigation purposes, leading to harm to the ecosystem balance and public health if populations are faced with potable water shortages. The only part of the "ethical fashion" definition that Joergens (2006) provides that is clear and makes good sense is the concept of fair-trade principles based on sweatshop-free production. Important, and much discussed over a decade after Joergens offered her definition, in light of the factory collapse tragedy in Bangladesh, that clarification has become the main focus of all ethical fashion campaigns. The problem is that the regulatory structures to reach a sweatshop-free production reality are absent. Their creation is hindered by the lack of coherent international regulatory standards for tolerable wages and working conditions.

When most production happens in underdeveloped nations with underdeveloped institutions, regulation and compliance are challenged by institutional inadequacy. Additionally, there is a lack of interpretation of what sweatshop working conditions mean in different cultural contexts. The problem is that its definition is applied in a Western ethnocentric understanding of what constitutes labor abuse, but there is a lack of agreement with the proposed standards when their interpretation is handed to decision makers in developing nations. What a Westerner considers sweatshop conditions, most non-Westerners see as normal working conditions.

As humans, we are most naturally impassioned by the suffering of other humans, and therefore labor policies for mitigating safety and occupational hazards in garment manufacturing can take center stage in the social responsibility discussion. The problem is that they overshadow the other issues that have a more direct ecological impact. It is an even bigger problem because of the advertising strategies in garment promotion. They are based on celebrity models, and those celebrity models are chosen for their social justice image. In the process, the true message of corporate social responsibility in production is represented by the impact on workers only, leaving out the ecological component.

When H&M uses as its face the rock star Madonna—a celebrity personality known for her humanitarian efforts in developing countries and support for environmental and social causes—it sends the message that its products are manufactured according to the ideals of the celebrity. The customer would not think that Madonna would lend her face to a company that is involved in humanitarian and ecologically irresponsible production. In a world of Twitter, where information is transmitted in 140 characters, when the face of a retailer has sustainability cachet, the average customer would edit his or her behavior down to that cachet.

Epilogue

WHAT CAN YOU DO?

1. Send Madonna, David Beckham, and Gisele Bundchen's PR teams a copy of this book. Or post a link to it on their social media sites.
2. If you are reading this book, chances are that you are a part of a demographic that can afford not to purchase any apparel products for a while. Try not to buy clothes for a year.
3. Do not buy clothes at Target, Walmart, Costco, or any large box stores.
4. If you are a parent, do not outsource the parenting of your child to "the mall." Spend time with your children and their friends in venues free from impulse-buying stimulation.
5. Be aware of the marketing tactics in retail overstimulation, and do not succumb to them.
6. Educate your children about impulse buying, marketing manipulation, and overconsumption.
7. And most importantly, never set foot in H&M, Zara, Forever 21, or any other fast-fashion retailer.

References

Abdelnour, R., and W. Peterson. (2007). The WTO decision on U.S. cotton policy. Institute of Agriculture and Natural Resources. Department of Agricultural Economics, University of Nebraska–Lincoln, Lincoln. Available at: http://www.agecon.unl.edu/Cornhuskereconomics.html.

Abecassis-Moedas, C. (2006). Integrating design and retail in the clothing value chain: An empirical study of the organisation of design. *International Journal of Operations & Production Management* 26(4): 412–428.

Abimbola, O. (2012). The international trade in secondhand clothing: Managing information asymmetry between West African and British traders. *Textile: The Journal of Cloth & Culture* 10(2): 184–199.

Adams, G. F. (2008). Globalization: From Heckscher–Ohlin to the new economic geography. *World Economics* 9(2): 153–174.

Adams, G. F. (2013). *Globalization; Today and Tomorrow*. London: Routledge.

Adhikari, R., and Y. Yamamoto. (2008). Chapter 1: Textile and clothing industry: Adjusting to the post-quota world. In Mikic, M., Zengpei, X., and Bonapace, T. (eds.), *Unveiling Protectionism: Regional Responses to Remaining Barriers in the Textile and Clothing Trade*. New York: United Nations ESCAP.

Ahmed, S. A., and M. Ali. (2004). Partnerships for solid waste management in developing countries: Linking theories to realities. *Habitat International* 28(3): 467–479.

Ait-Sahalia, Y., J. A. Parker, and M. Yogo. (2004). Luxury goods and the equity premium. *The Journal of Finance* 59(6): 2959–3004.

Aizenman, J., and E. Brooks. (2008). Globalization and taste convergence: The cases of wine and beer. *Review of International Economics* 16(2): 217–233.

Akamatsu, K. (1961). A theory of unbalanced growth in the world economy. *Hamburg: Weltwirtschaftliches Archiv* 86: 196–217.

Akamatsu, K. (1962). A historical pattern of economic growth in developing countries. *The Developing Economies* 1(1): 3–25.

Akram, A., D. Merunka, and M. Shakaib Akram. (2011). Perceived brand globalness in emerging markets and the moderating role of consumer ethnocentrism. *International Journal of Emerging Markets* 6(4): 291–303.

Ali, A. H. (2011). Power of social media in developing nations: New tools for closing the global digital divide and beyond. *Harvard Human Rights Journal* 24: 185–219.

Allenby, G. M., L. Jen, and R. P. Leone. (1996). Economic trends and being trendy: The influence of consumer confidence on retail fashion sales. *Journal of Business & Economic Statistics* 14(1): 103–111.

Andersen, L. P., B. Tufle, J. Rasmussen, and K. Chan. (2008). The tweens market and responses to advertising in Denmark and Hong Kong. *Young Consumers: Insight and Ideas for Responsible Marketers* 9(3): 189–200.

Anderson, A. (2011). Sources, media, and modes of climate change communication: The role of celebrities. *Wiley Interdisciplinary Reviews: Climate Change* 2(4): 535–546.

Anderson, A. (2013). Together we can save the Arctic: Celebrity advocacy and the Rio earth summit 2012. *Celebrity Studies* 4(3): 339–352.

Anderson, K., and E. Valenzuela. (2007). The World Trade Organization's Doha cotton initiative: A tale of two issues. *The World Economy* 30(8): 1281–1304.

Anguelov, N. (2014). *Policy and Political Theory in Trade Practice: Multinational Corporations and Global Governments.* New York: Palgrave Macmillan.

Anguelov, N. (2015). *Economic Sanctions versus Soft Power: Lessons from Myanmar, North Korea and the Middle East.* New York: Palgrave Macmillan.

Anholt, S. (2003). *Brand New Justice: The Upside of Global Branding.* Oxford: Butterworth-Heinemann.

Applebaum, W. (1951). Studying customer behavior in retail stores. *The Journal of Marketing* 16(2): 172–178.

Asche, H., and M. Schüller. (2008). China's engagement in Africa—Opportunities and risks for development. African Department, Economic Affairs, Eschborn: Deutsche Gesellschaft für Technische Zusammenarbeit (GTZ). Available at: http://s3.amazonaws.com/zanran_storage/www2.gtz.de/ContentPages/19176160.pdf.

Aslund, A. (2012). *How Capitalism Was Built: The Transformation of Central and Eastern Europe, Russia, The Caucasus, and Central Asia.* Cambridge: Cambridge University Press.

Aspers, P. (2012). *Markets in Fashion: A Phenomenological Approach.* London: Routledge.

Audet, D. (2004). Structural adjustment in textile and clothing in the post-ATC trading environment, OECD Trade Policy Papers 4. Available at: http://www.oecd.org/dataoecd/28/2/33672979.pdf.

Azmeh, S., and K. Nadvi. (2013). "Greater Chinese" global production networks in the middle east: The rise of the Jordanian garment industry. *Development and Change* 44(6): 1317–1340.

Baden, S., and C. Barber. (2005). The impact of the second-hand clothing trade on developing countries. Report prepared for Oxfam International, Oxford, UK.

Bae, J. H., and T. May-Plumlee. (2005). Customer focused textile and apparel manufacturing systems: Toward an effective e-commerce model. *Journal of Textile and Apparel, Technology and Management* 4(4): 1–19.

Baffes, J. (2005). The "cotton problem". *The World Bank Research Observer* 20(1): 109–144.

Baffes, J. (2007). Distortions to cotton sector incentives in West and Central Africa. Agricultural Distortion Working Paper #50. Washington, DC: World Bank.

Baffes, J. (2009). Chapter 18: Benin, Burkina Faso, Chad, Mali, and Togo. In Anderson, K., and Masters, W. A. (eds.), *Distortions to Agricultural Incentives in Africa.* Washington, DC: World Bank Publications.

Baffes, J. (2010). *Learning from "The Cotton Problem": Settling Trade Disputes.* Washington, DC: Carnegies Endowment for International Peace.

Baffes, J. (2011). Cotton subsidies, the WTO, and the "Cotton Problem". *The World Economy* 34(9): 1534–1556.

Bakan, J. (2012). *Childhood under Siege: How Big Business Targets Your Children.* New York: Simon and Schuster.

Bakewell, C., and V.-W. Mitchell. (2003). Generation Y female consumer decision-making styles. *International Journal of Retail & Distribution Management* 31(2): 95–106.

Balassa, B. (2012). *The Theory of Economic Integration* (Routledge Revivals). London: Routledge.

Baltagi, B. H., P. Egger, and M. Pfaffermayr. (2008). Estimating regional trade agreement effects on FDI in an interdependent world. *Journal of Econometrics* 145(1): 194–208.

Banuri, T. (1998). Pakistan: Environmental impact of cotton production and trade. Paper prepared for the United Nations Environment Programme.

Barker, C. (1999). *Television, Globalization and Cultural Identities*. Buckingham: Open University Press.

Barns, L., and G. Lea-Greenway. (2006). Fast fashioning the supply chain: Shaping the research agenda. *Journal of Fashion Marketing and Management* 10(3): 259–271.

Bateman, I. J., and K. G. Willis. (2001). *Valuing Environmental Preferences: Theory and Practice of the Contingent Valuation Method in the US, EU, and Developing Countries*. Oxford: Oxford University Press.

Bayley, G., and C. Nancarrow. (1998). Impulse purchasing: A qualitative exploration of the phenomenon. *Qualitative Market Research: An International Journal* 1(2): 99–114.

Beard, N. D. (2008). The branding of ethical fashion and the consumer: A luxury niche or mass-market reality? *Fashion Theory: The Journal of Dress, Body & Culture* 12(4): 447–468.

Bellman, L. M., I. Teich, and S. D. Clark. (2009). Fashion accessory buying intentions among female millennials. *Review of Business* 30(1): 46–57.

Bergadaà, M. (2007). Children and business: Pluralistic ethics of marketers. *Society and Business Review* 2(1): 53–73.

Bertoli, G., and R. Rosciniti (eds.). (2012). *International Marketing and the Country of Origin Effect: The Global Impact of "Made in Italy"*. Northampton, MA: Edward Elgar Publishing.

Bhaduri, G., and J. E. Ha-Brookshire. (2011). Do transparent business practices pay? Exploration of transparency and consumer purchase intention. *Clothing and Textiles Research Journal* 29(2): 135–149.

Bhardwaj, V., and A. Fairhurst. (2010). Fast fashion: Response to changes in the fashion industry. *The International Review of Retail, Distribution and Consumer Research* 20(1): 165–173.

Bhargava, H. K. (2012). Retailer-driven product bundling in a distribution channel. *Marketing Science* 31(6): 1014–1021.

Bijmolt, T., H. J. van Heerde, and R. G. M. Pieters. (2005). New empirical generalizations on the determinants of price elasticity. *Journal of Marketing Research* 42(2): 141–156.

Bikhchandani, S., D. Hirshleifer, and I. Welch. (1992). A theory of fads, fashion, custom, and cultural change as informational cascades. *Journal of Political Economy* 100(5): 992–1026.

Birnbaum, D. (2005). *Birnbaum's Global Guide to Winning the Great Garment War* (5th edn). Hong Kong: Third Horizon Press.

Birnbaum, D. (2008). *Crisis in the 21st Century Garment Industry and Breakthrough Unified Strategy*. New York: The Fashion Index Inc.

Blackmon, L. (2007). The devil wears Prado: A look at the design piracy prohibition act and the extension of copyright protection to the world of fashion. *Pepperdine Law Review* 35(1): 107–159.

Borders, M., and H. Sterling Burnett. (2006). Farm subsidies: Devastating the world's poor and the environment. National Center for Policy Analysis, Brief Analysis No. 547.

Bornstein, M. H., and R. H. Bradley (eds.). (2014). *Socioeconomic Status, Parenting, and Child Development*. Abingdon, UK: Routledge.

Brautignam, D. (2008). Flying geese or "hidden dragon"? Chinese business and African industrial development. In Alden, C., Large, D., and De Oliveira, R. S. (eds.), *China Returns to Africa: A Rising Power and a Continent Embrace*. New York: Columbia University Press.

Bridges, S., K. L. Keller, and S. Sood. (2000). Communication strategies for brand extensions: Enhancing perceived fit by establishing explanatory links. *Journal of Advertising* 29(4): 1–11.

Bridson, K., and J. Evans. (2004). The secret to a fashion advantage is brand orientation. *International Journal of Retail & Distribution Management* 32(8): 403–411.

Brockington, D. (2008). Powerful environmentalisms: Conservation, celebrity and capitalism. *Media, Culture, and Society* 30(4): 551–568.

Brooks, A. (2012). Riches from rags or persistent poverty? The working lives of second-hand clothing vendors in Maputo, Mozambique. *Textile: The Journal of Cloth and Culture* 10(2): 222–237.

Brooks, A. (2013). Stretching global production networks: The international second-hand clothing trade. *Geoforum* 44: 10–22.

Brooks, A., and D. Simon. (2012). Unravelling the relationships between used clothing imports and the decline of African clothing industries. *Development and Change* 43(6): 1265–1290.

Bruce, M., and L. Daly. (2006). Buyer behavior for fast fashion. *Journal of Fashion Marketing and Management* 10(3): 329–344.

Bruce, M., and L. Daly. (2011). Adding value: Challenges for UK apparel supply chain management—A review. *Production Planning & Control* 22(3): 210–220.

Bruce, M., L. Daly, and N. Towers. (2004). Lean or agile: A solution for supply chain management in the textiles and clothing industry? *International Journal of Operations & Production Management* 24(2): 151–170.

Buckley, P. J., and P. N. Ghauri. (2004). Globalization, economic geography and the strategy of multinational enterprise. *Journal of International Business Studies* 35(2): 81–98.

Bulgaria Fact Sheet 2004. (2004) Textile and clothing sector. Bulgarian government Investment Promotion Agency "Invest Bulgaria." Available at: http://www.bcci.bg/analytica/2005/attach/FS%202004%20textile%20and%20clothing%20sector.pdf.

Busse, M. (2010). On the growth performance of sub-Saharan African countries. *The Estey Centre Journal of International Policy and Trade* 11(2): 384–402.

Byun, S.-E., and B. Sternquist. (2011). Fast fashion and in-store hoarding: The drivers, moderator, and consequences. *Clothing and Textiles Research Journal* 29(3): 187–201.

Cachon, G. P., and R. Swinney. (2011). The value of fast fashion: Quick response, enhanced design, and strategic consumer behavior. *Management Science* 57(4): 778–795.

Carroll, A. (2009). Brand communications in fashion categories using celebrity endorsement. *Journal of Brand Management* 17(2): 146–158.

Cassidy, T. D., and H. van Schijndel. (2011). Youth identity ownership from a fashion marketing perspective. *Journal of Fashion Marketing and Management* 15(2): 163–177.

Cavusgil, T., and G. Knight. (2009). *Born Global Firms: A New International Enterprise.* New York: Business Expert Press.

Cervellon, M.-C. (2012). Victoria's dirty secrets: Effectiveness of green not-for-profit messages targeting brands. *Journal of Advertising* 41(4): 133–145.

Chase-Dunn, C. (1975). The effects of international economic dependence on development and inequality: A cross-national study. *American Sociological Review* 40: 720–738.

Chaturvedi, S., and G. Nagpal. (2003). WTO and product-related environmental standards: Emerging issues and policy options. *Economic and Political Weekly* 38(1): 66–74.

Chen, A. C.-H., R. Ya-Hui Chang, A. Besherat, and D. W. Baack. (2013). Who benefits from multiple brand celebrity endorsements? An experimental investigation. *Psychology & Marketing* 30(10): 850–860.

Chen, W., and B. Wellman. (2004). The global digital divide—within and between countries. *IT & Society* 1(7): 39–45.

Chen-Yu, J. H., and Y. K. Seock. (2002). Adolescents' clothing purchase motivations, information sources, and store selection criteria: A comparison of male/female and impulse/nonimpulse shoppers. *Family and Consumer Sciences Research Journal* 31(1): 50–77.

Choi, S. M., W.-N. Lee, and H.-J. Kim. (2005). Lessons from the rich and famous: A cross-cultural comparison of celebrity endorsement in advertising. *Journal of Advertising* 34(2): 85–98.

Choi, T.-M., N. Liu, S.-C. Liu, J. Mak, and Y.-T. To. (2010). Fast fashion brand extensions: An empirical study of consumer preferences. *Journal of Brand Management* 17(7): 472–487.

Chowdhury, A., and G. Mavrotas. (2006). FDI and growth: What causes what? *World Economy* 29(1): 9–19.

Christopher, M., R. Lowson, and H. Peck. (2004). Creating agile supply chains in the fashion industry. *International Journal of Retail & Distribution Management* 32(8): 367–376.

Christopher, M., and D. Towill. (2001). An integrated model for the design of agile supply chains. *International Journal of Physical Distribution & Logistics Management* 31(4): 235–246.

Chua, B. H. (2003). *Life is Not Complete without Shopping: Consumption Culture in Singapore*. Singapore: NUS Press.

Claudio, L. (2007). Waste couture: Environmental impact of the clothing industry. *Environmental Health Perspectives* 115(9): A449–A454.

Cline, E. L. (2012). *Overdressed: The Shockingly High Cost of Cheap Fashion*. New York: Penguin.

Clover, V. T. (1950). Relative importance of impulse-buying in retail stores. *The Journal of Marketing* 15(1): 66–70.

Cody, K. (2012). 'No Longer, But Not Yet': Tweens and the mediating of threshold selves through liminal consumption. *Journal of Consumer Culture* 12(1): 41–65.

Collins, R. (2002). *Media and Identity in Contemporary Europe: Consequences of Global Convergence*. Chicago: Intellect Books.

Colucci, M., and D. Scarpi. (2013). Generation Y: Evidences from the fast-fashion market and implications for targeting. *Journal of Business Theory and Practice* 1(1): 1–7.

Comstock, G., and E. Scharrer. (2010). *Media and the American Child*. Burlington, MA: Academic Press.

Craig, T. C., and R. King. (2002). *Global Goes Local: Popular Culture in Asia*. Vancouver: University of British Columbia Press.

Crofton, S. O., and L. G. Dopico. (2007). Zara-Inditex and the growth of fast fashion. *Essays in Economic & Business History* 25: 41–54.

Crothers, L. (2012). *Globalization and American Popular Culture*. Lanham, MD: Rowman & Littlefield.

Csikszentmihalyi, M., and B. Schneider. (2000). *Becoming Adult: How Teenagers Prepare for the World of Work*. New York: Basic Books.

Davis, F. (1992). *Fashion, Culture and Identity*. Chicago, IL: University of Chicago Press.

Debreu, G. (1987). *Theory of Value: An Axiomatic Analysis of Economic Equilibrium* (Vol. 17). Yale, CT: Yale University Press.

De Chernatony, L., C. Halliburton, and R. Bernath. (1995). International branding: Demand- or supply-driven opportunity? *International Marketing Review* 12(2): 9–21.

Delpeuch, C. (2007). EU and US safeguards against Chinese textile exports: What consequences for West African cotton-producing countries? MPRA Paper # 2317. Munich, Germany: University Library of Munich.

Dem, S. B., J. M. Cobb, and D. E. Mullins. (2007). Pesticide residues in soil and water from four cotton growing areas of Mali, West Africa. *Journal of Agricultural, Food and Environmental Sciences* 1(1): 16–28.

De Mooij, M. (2013). *Global Marketing and Advertising: Understanding Cultural Paradoxes.* Los Angeles: Sage Publications.

Dewsnap, B., and C. Hart. (2004). Category management: A new approach for fashion marketing? *European Journal of Marketing* 38(7): 809–834.

Dholakia, U. M., and D. Talukdar. (2004). How social influence affects consumption trends in emerging markets: An empirical investigation of the consumption convergence hypothesis. *Psychology & Marketing* 21(10): 775–797.

Dick, A. S., and K. Basu. (1994). Customer loyalty: Toward an integrated conceptual framework. *Journal of the Academy of Marketing Science* 22(2): 99–113.

Diebäcker, M. (2000). Environmental and social benchmarking for industrial processes in developing countries: A pilot project for the textile industry in India, Indonesia and Zimbabwe. *Integrated Manufacturing Systems* 11(7): 491–500.

Dimitrova, I., J. Kosturkov, and A. Vatralova. (1998). Industrial surface water pollution in the region of Devnya, Bulgaria. *Water Science and Technology* 37(8): 45–53.

Dinesh, G. P. (2012). Advertising and promotion campaigns as branding tools on teenagers. *Asian Journal of Research in Marketing* 1(2): 6–13.

D'Innocenzio, A. (2012). J.C. Penney says "No Sale": Cuts all prices, all the time to simplify bargain hunting. The Associated Press, January 25.

Dixit, A. K., and J. E. Stiglitz. (1977). Monopolistic competition and optimum product diversity. *The American Economic Review* 67(3): 297–308.

Domzal, T., and J. B. Kernan. (1993). Mirror, mirror: Some postmodern reflections on global advertising. *The Journal of Advertising* 22(4): 1–20.

Donadio, R. (2010). Chinese remake the "Made in Italy" fashion label. *The New York Times*, September 12.

Doss, S. (2011). The transference of brand attitude: The effect on the celebrity endorser. *Journal of Management and Marketing Research* 7(1): 58–70.

Dotson, M. J., and E. M. Hyatt. (2005). Major influence factors in children's consumer socialization. *Journal of Consumer Marketing* 22(1): 35–42.

Douglas, S. P., C. S. Craig, and E. J. Nijssen. (2001). Integrating branding strategy across markets: Building international brand architecture. *Journal of International Marketing* 9(2): 97–114.

Doyle, S. A., C. M. Moore, and L. Morgan. (2006). Supplier management in fast moving fashion retailing. *Journal of Fashion Marketing and Management* 10(3): 272–281.

Dresner, S. (2009). *The Principles of Sustainability.* London: Earthscan.

D'Souza, C. (2015). Marketing challenges for an eco-fashion brand: A case study. *Fashion Theory: The Journal of Dress, Body & Culture* 19(1): 67–82.

Du, S., C. B. Bhattacharya, and S. Sen. (2010). Maximizing business returns to corporate social responsibility (CSR): The role of CSR communication. *International Journal of Management Reviews* 12(1): 8–19.

Dunning, J. H. (2013). *International Production and the Multinational Enterprise.* Abingdon, UK: Routledge.

Easterly, W. (2005). What did structural adjustment adjust?: The association of policies and growth with repeated IMF and World Bank adjustment loans. *Journal of Development Economics* 76(1): 1–22.

Economist. (2014). A troubling trajectory: Fears are growing that trade's share of the world's GDP has peaked. But that is far from certain, December 14.

Elbehri, A. (2004). MFA quota removal and global textile and cotton trade: Estimating quota trade restrictiveness and quantifying post-MFA trade patterns. Prepared for the 7th annual conference on global economic analysis, Washington, DC.

Elliott, R., and C. Leonard. (2004). Peer pressure and poverty: Exploring fashion brands and consumption symbolism among children of the 'British Poor'. *Journal of Consumer Behaviour* 3(4): 347–359.

Elliott, R., and K. Shimamoto. (2008). Are ASEAN countries pollution havens for Japanese pollution-intensive industries? *The World Economy* 31(2): 236–254.

Ellis, J. L., V. A. McCracken, and N. Skuza. (2012). Insights into willingness to pay for organic cotton apparel. *Journal of Fashion Marketing and Management* 16(3): 290–305.

Epstein, D., S. O'Halloran, and A. L. Widsten. (2009). Implementing the agreement: Partisan politics and WTO dispute settlement. *Frontiers of Economics and Globalization* 6(5): 121–138.

Erdogan, B. Z. (1999). Celebrity endorsement: A literature review. *Journal of Marketing Management* 15(4): 291–314.

Eyhorn, F. (2007). *Organic Farming for Sustainable Livelihoods in Developing Countries?: The Case of Cotton in India.* Zurich: vdf Hochschulverlag AG.

Farrant, L., S. I. Olsen, and A. Wangel. (2010). Environmental benefits from reusing clothes. *The International Journal of Life Cycle Assessment* 15(7): 726–736.

Ferrigno, S., and A. Lizarraga. (2009). Components of a sustainable cotton production system: Perspective from the organic cotton experience. *ICAC Recorder* March: 13–23.

Ferrigno, S., S. G. Ratter, P. Ton, D. S. Vodouhê, S. Williamson, and J. Wilson. (2005). *Organic Cotton: A New Development Path for African Smallholders?* London: International Institute for Environment and Development.

Fiore, A. M. (2010). *Understanding Aesthetics for the Merchandising and Design Professional.* London: Bloomsbury Publishing.

Fortucci, P. (2002) The contribution of cotton to economy and food security in developing countries. Note presented at the Conference "Cotton and Global Trade Negotiations" sponsored by the World Bank and ICAC, Washington, DC: The World Bank.

Forza, C., and A. Vinelli. (2000). Time compression in production and distribution within the textile-apparel chain. *Integrated Manufacturing Systems* 11(2): 138–146.

Freidman, T. (2005). *The World is Flat.* New York: Farrar, Straus and Giroux.

Furnham, A. (1999). The saving and spending habits of young people. *Journal of Economic Psychology* 20(6): 677–697.

Gabrielli, V., I. Baghi, and V. Codeluppi. (2013). Consumption practices of fast fashion products: A consumer-based approach. *Journal of Fashion Marketing and Management* 17(2): 206–224.

Gaimster, J. (2012). The changing landscape of fashion forecasting. *International Journal of Fashion Design, Technology and Education* 5(3): 169–178.

Gam, H. J. (2011). Are fashion-conscious consumers more likely to adopt eco-friendly clothing? *Journal of Fashion Marketing and Management* 15(2): 178–193.

Gammoh, B., K. E. Voss, and X. Fang. (2010). Multiple brand alliances: A portfolio diversification perspective. *Journal of Product and Brand Management* 19(1): 27–33.

Ganesan, N. (1999). *Bilateral Tensions in Post-Cold War ASEAN.* Singapore: Institute of Southeast Asian Studies.

Ganesan, N. (2001). Thailand's relations with Malaysia and Myanmar in post-cold war Southeast Asia. *Japanese Journal of Political Science* 2(1): 127–146.

Ger, G., and R. W. Belk. (1996). I'd like to buy the world a coke: Consumptionscapes of the "less affluent world". *Journal of Consumer Policy* 19(3): 271–304.

Gereffi, G. (1999). International trade and industrial upgrading in the apparel commodity chain. *Journal of International Economics* 48(1): 37–70.

Gereffi, G. (2001). Shifting governance structures in global commodity chains, with special reference to the internet. *American Behavioral Scientist* 44(10): 1616–1637.

Gereffi, G. (2006). *The New Offshoring of Jobs and Global Development.* Geneva: International Labour Organization.

Gereffi, G., and S. Frederick. (2010). The global apparel value chain, trade and the crisis: Challenges and opportunities for developing countries. World Bank Policy Research. Working Paper #5281. Washington, DC: The World Bank.

Gereffi, G., and O. Memedovic. (2003). *The Global Apparel Value Chain: What Prospects for Upgrading by Developing Countries.* Vienna: United Nations Industrial Development Organization.

Gibbon, P. (2003). The Africa growth and opportunity act and the global commodity chain for clothing. *World Development* 31(11): 1809–1827.

Giertz-Mårtenson, I. (2012). H&M—Documenting the story of one of the world's largest fashion retailers. *Business History* 54(1): 108–115.

Grannis, K. (2014). Demand for clothing, supplies, electronics drives school/college spending up. Available at: nrf.com, July 17.

Grant, I. C. (2004). Communicating with young people through the eyes of marketing practitioners. *Journal of Marketing Management* 20(5–6): 591–606.

Gray, R. (2006). Social, environmental and sustainability reporting and organisational value creation?: Whose value? Whose creation? *Accounting, Auditing & Accountability Journal* 19(6): 793–819.

Greene, W. H. (2008). *Econometric Analysis* (6th edn). Upper Saddle River, NJ: Pearson/ Prentice Hall.

Greer, L., S. E. Keane, and Z. Lin. (2010). *NRDC's Ten Best Practices for Textile Mills to Save Money and Reduce Pollution.* National Resource Defense Council. Available at: http://www.nrdc.org/international/cleanbydesign/files/rsifullguide.pdf.

Grimes, P., and J. Kentor. (2003). Exporting the greenhouse: Foreign capital penetration and CO_2 emissions 1980–1996. *Journal of World-Systems Research* 9(2): 261–275.

Guillén, M. F., and S. L. Suárez. (2005). Explaining the global digital divide: Economic, political and sociological drivers of cross-national internet use. *Social Forces* 84(2): 681–708.

HaeJung, K. (2012). The dimensionality of fashion-brand experience. *Journal of Fashion Marketing and Management* 16(4): 418–441.

Haggblade, S. (1990). The flip side of fashion: Used clothing exports to the third world. *The Journal of Development Studies* 26(3): 505–521.

Haig, M. (2006). *Brand Royalty: How the World's Top 100 Brands Thrive & Survive.* London: Kogan Page Publishers.

Hale, A., and J. Wills. (2005). *Threads of Labour: Garment Industry Supply Chains from the Workers' Perspective*. Malden, MA: Blackwell.

Hancock, J. (2009). *Brand/Story: Ralph, Vera, Johnny, Billy, and Other Adventures in Fashion Branding*. London: Bloomsbury Publishing.

Hansen, K. T. (1999). Second-hand clothing encounters in Zambia: Global discourses, western commodities, and local histories. *Africa* 69(3): 343–365.

Hansen, K. T. (2000). Other people's clothes? The international second-hand clothing trade and dress practices in Zambia. *Fashion Theory: The Journal of Dress, Body & Culture* 4(3): 245–274.

Hansen, K. T. (2004). Helping or hindering? Controversies around the international second-hand clothing trade. *Anthropology Today* 20(4): 3–9.

Hansen, S. (2012). How Zara grew into the world's largest fashion retailer. *The New York Times Magazine* November 9.

Hart, D., R. Atkins, P. Markey, and J. Youniss. (2004). Youth bulges in communities: The effects of age structure on adolescent civic knowledge and civic participation. *Psychological Science* 15(9): 591–597.

Hartmann, P., and V. A. Ibanez. (2006). Green value added. *Marketing Intelligence & Planning* 24(7): 673–680.

Hassett, K. (ed.). (2012). *Rethinking Competitiveness*. Lanham, MD: Rowman & Littlefield.

Hayes, S. G., and N. Jones. (2006). Fast fashion: A financial snapshot. *Journal of Fashion Marketing and Management* 10(3): 282–300.

Haytko, D. L., and J. Baker. (2004). It's all at the mall: Exploring adolescent girls' experiences. *Journal of Retailing* 80(1): 67–83.

Heinisch, E. L. (2006). West Africa versus the United States on cotton subsidies: How, why and what next? *The Journal of Modern African Studies* 44(2): 251–274.

Heller, L. (2014). The real cost of fast fashion. Forbes.com, 28 April. Available at: http://www.forbes.com/sites/lauraheller/2014/04/28/the-real-cost-of-fast-fashion/

Henry, R. K., Z. Yongsheng, and D. Jun. (2006). Municipal solid waste management challenges in developing countries—Kenyan case study. *Waste Management* 26(1): 92–100.

Hicks, J. (1970). Elasticity of substitution again: Substitutes and complements. *Oxford Economic Papers* 22(3): 289–296.

Hill, D. (2011). Guest worker programs are no fix for our broken immigration system: Evidence from the Northern Mariana Islands. *New Mexico Law Review* 41(1): 131–191.

Hill, J., and H.-H. Lee. (2012). Young generation Y consumers' perceptions of sustainability in the apparel industry. *Journal of Fashion Marketing and Management* 16(4): 477–491.

Hutson, A., M. Biravadolu, and G. Gereffi. (2005). Value chains for the U.S. cotton industry. Report prepared for Oxfam America, Boston, MA.

Hyllegard, K. H., R.-N. Yan, J. P. Ogle, and K.-H. Lee. (2012). Socially responsible labeling: The impact of hang tags on consumers' attitudes and patronage intentions toward an apparel brand. *Clothing and Textiles Research Journal* 30(1): 51–66.

Ibrahim, N. A., N. M. Moneim, H. Adbel, E. S. Abdel, and M. M. Hosni. (2008). Pollution prevention of cotton-cone reactive dyeing. *Journal of Cleaner Production* 16(12): 1321–1326.

Jacobs, B. (2011). A dragon and a dove? A comparative overview of Chinese and European trade relations with Sub-Saharan Africa. *Journal of Current Chinese Affairs* 40(4): 17–60.

James, D. (2005). Guilty through association: Brand association transfer to brand alliances. *Journal of Consumer Marketing* 22(1): 14–24.

Janiszewski, C., and S. M. J. Van Osselaer. (2000). A connectionist model of brand–quality associations. *Journal of Marketing Research* 37(3): 331–350.

Jansson, J., and D. Power. (2010). Fashioning a global city: Global city brand channels in the fashion and design industries. *Regional Studies* 44(7): 889–904.

Jaquette, J. S. (1997). Women in power: From tokenism to critical mass. *Foreign Policy* 108: 23–37.

Jauch, H., and R. Traub-Merz (eds.). (2006). *The Future of the Textile and Clothing Industry in Sub-Saharan Africa*. Bonn: Friedrich-Ebert-Stiftung.

Jeppesen, S., and M. W. Hansen. (2004). Environmental upgrading of third world enterprises through linkages to transnational corporations. Theoretical perspectives and preliminary evidence. *Business Strategy and the Environment* 13(4): 261–274.

Jewitt, S. (1995). Europe's "others"? Forestry policy and practices in colonial and postcolonial India. *Environment and Planning* D(13): 67–67.

Jiang, L., and R. Anupindi. (2010). Customer-driven vs. retailer-driven search: Channel performance and implications. *Manufacturing & Service Operations Management* 12(1): 102–119.

Jimenez, G. C., and B. Kolsun (eds.). (2014). *Fashion Law: A Guide for Designers, Fashion Executives, and Attorneys*. New York: Bloomsbury.

Joergens, C. (2006). Ethical fashion: Myth or future trend? *Journal of Fashion Marketing and Management* 10(3): 360–371.

Johanson, J., and J.-E. Vahlne. (1977). The internationalization process of the firm: A model of knowledge development and increasing foreign commitments. *Journal of International Business Studies* 8(1): 23–32.

Johnson, T., and J. Attmann. (2009). Compulsive buying in a product specific context: Clothing. *Journal of Fashion Marketing and Management* 13(3): 394–405.

Jones, M. A., K. E. Reynolds, S. Weun, and S. E. Beatty. (2003). The product-specific nature of impulse buying tendency. *Journal of Business Research* 56(7): 505–511.

Jones, V. C., and B. R. Williams. (2012). US trade and investment relations with Sub-Saharan Africa and the African growth and opportunity act. Congressional Research Service Report # RL31772. Washington, DC: US Library of Congress.

Joomun, G. (2006). The textile and clothing industry in Mauritius. In Jauch, H., and Traub-Merz, R. (eds.), *The Future of the Textile and Clothing Industry in Sub-Saharan Africa* (pp. 193–211). Bonn, Germany: Friedrich-Ebert-Stiftung.

Jorgenson, A. K. (2003). Consumption and environmental degradation: A cross-national analysis of the ecological footprint. *Social Problems* 50(3): 374–394.

Jorgenson, A. K. (2006). Unequal ecological exchange and environmental degradation: A theoretical proposition and cross-national study of deforestation, 1990–2000. *Rural Sociology* 71(4): 685–712.

Jorgenson, A. K. (2007). Does foreign investment harm the air we breathe and the water we drink? A cross-national study of carbon dioxide emissions and organic water pollution in less-developed countries, 1975–2000. *Organization and Environment* 20(2): 137–156.

Jorgenson, A. K. (2009). Political-economic integration, industrial pollution and human health: A panel study of less-developed countries, 1980–2000. *International Sociology* 24(1): 115–143.

Jovanovic, M. (2014). *International Economic Integration: Limits and Prospects*. London: Routledge.

Joy, A., J. F. Sherry, A. Venkatesh, J. Wang, and R. Chan. (2012). Fast fashion, sustainability, and the ethical appeal of luxury brands. *Fashion Theory: The Journal of Dress, Body & Culture* 16(3): 273–296.

Kahneman, D., and J. L. Knetsch. (1992). Valuing public goods: The purchase of moral satisfaction. *Journal of Environmental Economics and Management* 22(1): 57–70.

Kalla, S. M., and A. P. Arora. (2011). Impulse buying a literature review. *Global Business Review* 12(1): 145–157.

Kamau, P., D. McCormick, and N. Pinaud. (2009). The developmental impact of Asian drivers on Kenya with emphasis on textiles and clothing manufacturing. *The World Economy* 32(11): 1586–1612.

Kant, R. (2012). Textile dyeing industry: An environmental hazard. *Natural Science* 4(1): 22–26.

Kaplan, R. D. (2011). *Monsoon: The Indian Ocean and the Future of American Power*. New York: Random House.

Kapp, W. (1950). *The Social Costs of Private Enterprise*. New York: Schocken Books.

Kelley, C. A., D. Shen, and P. Tong. (2012). Developing marketing strategies to build stable trade relationships under WTO rules: The case of the entry of China's textile industry in the US. *Proceedings of the American Society of Business and Behavioral Sciences* 19(1): 495–506.

Kelting, K., and D. H. Rice. (2013). Should we hire David Beckham to endorse our brand? Contextual interference and consumer memory for brands in a celebrity's endorsement portfolio. *Psychology & Marketing* 30(7): 602–613.

Kentor, J., and P. Grimes. (2006). Foreign investment dependence and the environment: A global perspective. In Jorgenson, A. K., and Kick, E. (eds.), *Globalization and the Environment* (pp. 61–78). Leiden: Brill.

Khan, H. S., S. Ahmed, A. V. Evans, and M. Chadwick. (2009). Methodology for performance analysis of textile effluent treatment plants in Bangladesh. *Chemical Engineering Research Bulletin* 13: 61–66.

Khan, U. (2008). Liz Hurley's "safety pin" dress voted the greatest dress. *The Daily Telegraph*, October 9.

Kilcooley-O'Halloran, S. (2015). Gisele Bündchen covers March Vogue. *Vogue*, February 2.

Kim, H. (2012). The dimensionality of fashion-brand experience: Aligning consumer-based brand equity approach. *Journal of Fashion Marketing and Management* 16(4): 418–441.

Kinda, T. (2010). Investment climate and FDI in developing countries: Firm-level evidence. *World Development* 38(4): 498–513.

Klein, N. (1999). *No Logo*. Canada: Knopf.

Knight, G., and T. Cavusgil. (1996). The born global firm: A challenge to traditional internationalization theory. In Tamar, C., and Tage, M. (eds.), *Advances in International Marketing* (Vol. 8, pp. 11–26). Greenwich, CT: JAI Press.

Knight, G., and T. Cavusgil. (2004). Innovation, organizational capabilities, and the born-global firm. *Journal of International Business Studies* 35(2): 124–141.

Kopp, R. T., R. J. Eng, and D. J. Tigert. (1989). A competitive structure and segmentation analysis of the Chicago fashion market. *Journal of Retailing* 65(4): 496–515.

Kraidy, M. (2005). *Hybridity, or the Cultural Logic of Globalization*. Philadelphia, PA: Temple University Press.

Krugman, P. R., and E. Helpman. (1985). *Market Structure and Foreign Trade: Increasing Returns, Imperfect Competition and the International Economy*. Cambridge, MA: MIT Press.

Kulmala, M., N. Mesiranta, and P. Tuominen. (2013). Organic and amplified eWOM in consumer fashion blogs. *Journal of Fashion Marketing and Management* 17(1): 20–37.

Kumagai, S. (2008). A journey through the secret history of the flying geese model. IDE Discussion Paper No. 158, Institute of Developing Economies. Japan: JETRO.

Kumar, N. (1997). The revolution in retailing: From market driven to market driving. *Long Range Planning* 30(6): 830–835.

Kunz, G., and M. Garner. (2011). *Going Global: The Textile and Apparel Industry.* New York: Fairchild Books.

Laing, I. G. (1991). The impact of effluent regulations on the dyeing industry. *Review of Progress in Coloration and Related Topics* 21(1): 56–71.

Lal, R. (2004). Soil carbon sequestration impacts on global climate change and food security. *Science* 304(5677): 1623–1627.

Lazzeretti, L., S. R. Sedita, and A. Caloffi. (2014). Founders and disseminators of cluster research. *Economic Geography* 14(1): 21–43.

Lee, C.-G. (2009a). Foreign direct investment, pollution and economic growth: Evidence from Malaysia. *Applied Economics* 41(13): 1709–1716.

Lee, J. A., and J. J. Kacen. (2008). Cultural influences on consumer satisfaction with impulse and planned purchase decisions. *Journal of Business Research* 61(3): 265–272.

Lee, K. (2008a) Opportunities for green marketing: Young consumers. *Marketing Intelligence & Planning* 26(6): 573–586.

Lee, K. (2009b). Gender differences in Hong Kong adolescent consumers' green purchasing behavior. *Journal of Consumer Marketing* 26(2): 87–96.

Lee, M.-D. P. (2008b). A review of the theories of corporate social responsibility: Its evolutionary path and the road ahead. *International Journal of Management Reviews* 10(1): 53–73.

Lee, N., A. J. Broderick, and L. Chamberlain. (2007). What is "neuromarketing"? A discussion and agenda for future research. *International Journal of Psychophysiology* 63(2): 199–204.

Leiss, W. (2013). *Social Communication in Advertising: Consumption in the Mediated Marketplace.* London: Routledge.

Leslie, D. (1995). Global scan: The globalization of advertising agencies, concepts, and campaigns. *Economic Geography* 71(4): 402–426.

Leslie, D. (2002). Gender, retail employment and the clothing commodity chain. Gender, place and culture. *A Journal of Feminist Geography* 9(1): 61–76.

Lin, Y.-H., and C.-Y. Chen. (2012). Adolescents' impulse buying: Susceptibility to interpersonal influence and fear of negative evaluation. *Social Behavior and Personality: An International Journal* 40(3): 353–358.

Lloyd, A. E., and S. T. K. Luk. (2010). The Devil wears Prada or Zara: A revelation into customer perceived value of luxury and mass fashion brands. *Journal of Global Fashion Marketing* 1(3): 129–141.

Lopez, C., and Y. Fan. (2009). Internationalisation of the Spanish fashion brand Zara. *Journal of Fashion Marketing and Management* 13(2): 279–296.

Loughlin, C., and J. Barling. (2001). Young workers' work values, attitudes, and behaviours. *Journal of Occupational and Organizational Psychology* 74(4): 543–558.

Loureiro, M. L., J. McCluskey, and R. C. Mittelhammer. (2001). Assessing consumer preferences for organic, eco-labeled, and regular apples. *Journal of Agricultural and Resource Economics* 26(2): 404–416.

Low, G. S., and J. Mohr. (2000). Advertising vs. sales promotion: A brand management perspective. *Journal of Product & Brand Management* 9(6): 389–414.

Lu, S. (2012). China takes all? An empirical study on the impacts of quota elimination on world clothing trade from 2000 to 2009. *Journal of Fashion Marketing and Management* 16(3): 306–326.

MacCarthy, B. L., and P. G. S. A. Jayarathne. (2010) Fast fashion: Achieving global quick response (GQR) in the internationally dispersed clothing industry. In Cheng, E., and Tsan-Ming, C. (eds.), *Innovative Quick Response Programs in Logistics and Supply Chain Management* (pp. 37–60). Berlin/Heidelberg: Springer.

MacDonald, S., S. Pan, A. Somwaru, and F. Tuan. (2010). China's role in world cotton and textile markets: A joint computable general equilibrium/partial equilibrium approach. *Applied Economics* 42(7): 875–885.

Madhavaram, S. R., and D. A. Laverie. (2004). Exploring impulse purchasing on the internet. *Advances in Consumer Research* 31(1): 59–66.

Magrath, V., and H. McCormick. (2013a). Branding design elements of mobile fashion retail apps. *Journal of Fashion Marketing and Management* 17(1): 98–114.

Magrath, V., and H. McCormick. (2013b). Marketing design elements of mobile fashion retail apps. *Journal of Fashion Marketing and Management* 17(1): 115–134.

Mangieri, T. (2006). African cloth, export production, and secondhand clothing in Kenya. Working paper. Department of Geography, University of North Carolina at Chapel Hill.

Mantzavinos, C. (2001). *Individuals, Institutions, and Markets*. Cambridge: Cambridge University Press.

Markusen, A. (1996). Sticky places in slippery space: A typology of industrial districts. *Economic Geography* 72(3): 293–313.

Martin, M. F. (2007). US clothing and textile trade with China and the world: Trends since the end of quotas. Congressional Research Services Report # RL34106. Washington, DC: Library of Congress.

Martinez, L. A. (2012). The country-specific nature of apparel elasticities and impacts of the multi-fibre arrangement. Honors Projects Series, Paper # 49. St. Paul, MN: Macalester College.

Marwell, G., and P. Oliver. (1993). *The Critical Mass in Collective Action*. Cambridge: Cambridge University Press.

Matten, D., and J. Moon. (2008). "Implicit" and "explicit" CSR: A conceptual framework for a comparative understanding of corporate social responsibility. *Academy of Management Review* 33(2): 404–424.

Matthews, H., M. Taylor, B. Percy-Smith, and M. Limb. (2000). The unacceptable flaneur: The shopping mall as a teenage hangout. *Childhood* 7(3): 279–294.

Mattila, A. S., and J. Wirtz. (2008). The role of store environmental stimulation and social factors on impulse purchasing. *Journal of Services Marketing* 22(7): 562–567.

Mattioli, D., and K. Hudson. (2011). Gap to slash its store count. *The Wall Street Journal* October 14.

McCaughey, M., and M. D. Ayers (eds.). (2003). *Cyberactivism: Online Activism in Theory and Practice*. New York: Routledge.

McGrath, R. G. (2010). Business models: A discovery driven approach. *Long Range Planning* 43(2): 247–261.

McKinnon, R., and G. Schnabl. (2003). China: A stabilizing or deflationary influence in East Asia? The problem of conflicted virtue. Sanford Economics Working Paper #03007. Sanford, CA.

McMichael, P. (2004). *Development and Social Change: A Global Perspective* (3rd edn.). Thousand Oaks, CA: Pine Forge Press.

Melitz, M. J., and D. Trefler. (2012). Gains from trade when firms matter. *The Journal of Economic Perspectives* 26(2): 91–118.

Meyer, P. (2014). Brazil: Political and economic situation and US relations. Congressional Research Service, Report# RL33456. Washington, DC: Library of Congress.

Mikic, M., X. Zengpei, and T. Bonapace (eds.). (2008). *Unveiling Protectionism: Regional Responses to Remaining Barriers in the Textile and Clothing Trade*. New York: United Nations ESCAP.

Miller, C. M., S. H. McIntyre, and M. K. Mantrala. (1993). Toward formalizing fashion theory. *Journal of Marketing Research* 30(2): 142–157.

Minot, N. W., and L. Daniels. (2005). Impact of global cotton markets on rural poverty in Benin. *Agricultural Economics* 33(3): 453–466.

Miroux, A., and K. P. Sauvant (eds.). (2005). *TNCs and the Removal of Textiles and Clothing Quotas*. Geneva/New York: UNCTAD Current Studies on FDI and Development Series, United Nations Publication.

Mitchell, C., and J. Reid-Walsh (eds.). (2005). *Seven Going on Seventeen: Tween Studies in the Culture of Girlhood*. New York: Peter Lang Publishing.

Moeran, B. (2003). Celebrities and the name economy. In Dannhaeuser, N., and Werner, C. (eds.), *Anthropological Perspectives on Economic Development and Integration*. Research in Economic Anthropology (Vol. 22, pp. 299–321). Bradford, UK: Emerald Group Publishing.

Moore, C. M., and G. Birtwistle. (2004). The Burberry business model: Creating an international luxury fashion brand. *International Journal of Retail & Distribution Management* 32(8): 412–422.

Morgan, L. R., and G. Birtwistle. (2009). An investigation of young fashion consumers' disposal habits. *International Journal of Consumer Studies* 33(2): 190–198.

Moses, E. (2000). *The $100 Billion Allowance: Accessing the Global Teen Market*. Hoboken, NJ: Wiley.

Murray, S. (2005). Brand loyalties: Rethinking content within global corporate media. *Media, Culture & Society* 27(3): 415–435.

Nabar, M. (2011). Targets, interest rates, and household saving in urban China. Working Paper #11/223. Washington, DC: International Monetary Fund.

Naderi, I. (2013). Beyond the fad: A critical review of consumer fashion involvement. *International Journal of Consumer Studies* 37(1): 84–104.

Nenni, M. E., L. Giustiniano, and L. Pirolo. (2013). Demand forecasting in the fashion industry: A review. *International Journal of Engineering Business Management* 5(37): 1–6.

Nilkant, D. (2014). Buying behavior of teenagers in Bangalore: A special emphasis on apparels. *International Journal of Applied Services Marketing Perspectives* 3(3): 1194–1198.

Nimon, W., and J. Beghin. (1999). Are eco-labels valuable? Evidence from the apparel industry. *American Journal of Agricultural Economics* 81(4): 801–811.

Norris, L. (2010). *Recycling Indian Clothing: Global Contexts of Reuse and Value*. Bloomington, IN: Indiana University Press.

North, D. C. (1990). *Institutions, Institutional Change and Economic Performance*. Cambridge: Cambridge University Press.

O'Connor, C. (2014). These retailers involved in Bangladesh factory disaster have yet to compensate victims. Forbes.com, April 24.

O'Guinn, T., C. Allen, R. Semenik, and A. C. Scheinbaum. (2014). *Advertising and Integrated Brand Promotion*. Stamford, CT: Cengage Learning.

Okonkwo, U. (2007). *Luxury Fashion Branding: Trends, Tactics, Techniques*. New York: Palgrave Macmillan.

Ozsomer, A., and S. Altaras. (2008). Global brand purchase likelihood: A critical synthesis and an integrated conceptual framework. *Journal of International Marketing* 16(4): 1–28.

Pan, B., and R. Holland. (2006). A mass customised supply chain for the fashion system at the design-production interface. *Journal of Fashion Marketing and Management* 10(3): 134–159.

Pan, J., C. Chu, X. Zhao, Y. Cui, and T. Voituriez. (2008). *Global Cotton and Textile Product Chains: Identifying Challenges and Opportunities for China through a Global Commodity Chain Sustainability Analysis*. Winnipeg, Canada: International Institute for Sustainable Development.

Pan, S., C. Wang, and D. Hudson. (2010). Is Investment in agricultural research—A good substitute for price support in U.S. cotton? AAEA Selected Paper # 10810, Department of Agricultural and Applied Economics, Texas Tech University.

Pantsios, A. (2015). Pharrell and Al Gore announce "Live Earth Road to Paris". ecowatch. com, 21 January.

Park, E. J., E. Y. Kim, and J. C. Forney. (2006). A structural model of fashion-oriented impulse buying behavior. *Journal of Fashion Marketing and Management* 10(4): 433–446.

Park, E. J., E. Y. Kim, V. M. Funches, and W. Foxx. (2012). Apparel product attributes, web browsing, and e-impulse buying on shopping websites. *Journal of Business Research* 65(11): 1583–1589.

Park, H.-J., and N. J. Rabolt. (2009). Cultural value, consumption value, and global brand image: A cross-national study. *Psychology and Marketing* 26(8): 714–735.

Park, M., and S. J. Lennon. (2009). Brand name and promotion in online shopping contexts. *Journal of Fashion Marketing and Management* 13(2): 149–160.

Parker, R. S., C. M. Hermans, and A. D. Schaefer. (2004). Fashion consciousness of Chinese, Japanese and American teenagers. *Journal of Fashion Marketing and Management* 8(2): 176–186.

Parker, R. S., A. D. Schaefer, and C. M. Hermans. (2007). An investigation into teens' attitudes towards fast-food brands in general: A cross-cultural analysis. *Journal of Foodservice Business Research* 9(4): 25–40.

Peattie, K., and A. Crane. (2005). Green marketing: Legend, myth, farce or prophesy? *Qualitative Market Research: An International Journal* 8(4): 357–370.

Pells, R. (2011). *Modernist America: Art, Music, Movies, and the Globalization of American Culture*. New Haven, CT: Yale University Press.

Pentecost, R., and L. Andrews. (2010). Fashion retailing and the bottom line: The effects of generational cohorts, gender, fashion fanship, attitudes and impulse buying on fashion expenditure. *Journal of Retailing and Consumer Services* 17(1): 43–52.

Perfloff, J. (2010). *Microeconomics: Theory and Applications with Calculus* (2nd edn). Upper Saddle River, NJ: Pearson.

Pesendorfer, W. (1995). Design innovation and fashion cycles. *The American Economic Review* 85(4): 771–792.

Phan, M., R. Thomas, and K. Heine. (2011). Social media and luxury brand management: The case of Burberry. *Journal of Global Fashion Marketing* 2(4): 213–222.

Phau, I., and C.-C. Lo. (2004). Profiling fashion innovators: A study of self-concept, impulse buying and internet purchase intent. *Journal of Fashion Marketing and Management* 8(4): 399–411.

Phillips, B. J., and E. F. McQuarrie. (2011). Contesting the social impact of marketing: A re-characterization of women's fashion advertising. *Marketing Theory* 11(2): 99–126.

Pine, B. J., and J. H. Gilmore. (1999). *The Experience Economy: Work is Theatre & Every Business a Stage*. Boston, MA: Harvard Business Press.

Porter, M. E. (2000). Location, competition, and economic development: Local clusters in a global economy. *Economic Development Quarterly* 14(1): 15–34.

Porter, M. E., and M. R. Kramer. (2006). The link between competitive advantage and corporate social responsibility. *Harvard Business Review* 84(12): 78–92.

Presas, T. (2001). Interdependence and partnership: Building blocks to sustainable development. *Corporate Environmental Strategy* 8(3): 203–208.

Puig, F., B. García-Mora, and C. Santamaría. (2013). The influence of geographical concentration and structural characteristics on the survival chance of textile firms. *Journal of Fashion Marketing and Management* 17(1): 6–19.

Purvis, L., M. M. Naim, and D. Towill. (2013). Intermediation in agile global fashion supply chains. *International Journal of Engineering, Science and Technology* 5(2): 38–48.

Quart, A. (2008). *Branded: The Buying and Selling of Teenagers*. New York: Basic Books.

Quimby, F. (2013). Americanised, decolonised, globalised and federalised: The Northern Mariana Islands since 1978. *The Journal of Pacific History* 48(4): 464–483.

Qureshi, S. (2012). As the global digital divide narrows, who is being left behind? *Information Technology for Development* 18(4): 277–280.

Reichard, R. (2009). Textiles 2009: Some late bottoming out, but no miracles. textile-world.com, January/February.

Rice, J. (2007). Ecological unequal exchange: International trade and uneven utilization of environmental space in the world system. *Social Forces* 85(3): 1369–1392.

Richardson, J. (1996). Vertical integration and rapid response in fashion apparel. *Organization Science* 7(4): 400–412.

Rivoli, P. (2009). *The Travels of a T-Shirt in the Global Economy: An Economist Examines the Markets, Power, and Politics of World Trade*. London: Wiley.

Roberts, J. T., and B. C. Parks. (2007). *A Climate of Injustice: Global Inequality, North–South Politics, and Climate Policy*. Cambridge, MA: MIT Press.

Robinson, R. (2008). *Indonesia: The Rise of Capital*. Sheffield, UK: Equinox Press.

Robinson, T., and F. Doss (2011). Pre-purchase alternative evaluation: Prestige and imitation fashion products. *Journal of Fashion Marketing and Management* 15(3): 278–290.

Robson, A. J. (2002). Evolution and human nature. *Journal of Economic Perspectives* 16(2): 89–106.

Rohwedder, C., and K. Johnson. (2008). Pace-setting Zara seeks more speed to fight its rising cheap-chic rivals. *The Wall Street Journal*, February 20.

Rosenthal, E. (2007). Can polyester save the world? *The New York Times*, January 25.

Ross, J., and R. Harradine. (2011). Fashion value brands: The relationship between identity and image. *Journal of Fashion Marketing and Management* 15(3): 306–325.

Ryder, G. (2003). International Confederation of Free Trade Unions (ICFTU) Press Conference, September 12. World Trade Organization Fifth Ministerial Meeting, Cancún, Mexico.

Sagoff, M. (1998). Aggregation and deliberation in valuing environmental public goods: A look beyond contingent pricing. *Ecological Economics* 24(2): 213–230.

Saladino, M. P. (2008). The proliferation of product placement as a means of advertising communication. *Journal of International Business Ethics* 1(1): 100–106.

Schofer, E., and A. Hironaka. (2005). The effects of world society on environmental protection outcomes. *Social Forces* 84(1): 25–47.

Schor, J. (2004). *Born to Buy: The Commercialized Child and the New Consumer Culture*. New York: Simon and Schuster.

Schor, J. (2008). *The Overworked American: The Unexpected Decline of Leisure*. New York: Basic Books.

Schor, J. B. (1998). *The Overspent American: Why We Want What We Don't Need*. New York: Harper Collins.

Schröppel, C., and M. Nakajima. (2002). The changing interpretation of the flying geese model of economic development. *Japanstudien* 14: 203–236.

Schuiling, I., and J.-N. Kapferer. (2004). Executive insights: Real differences between local and international brands: Strategic implications for international marketers. *Journal of International Marketing* 12(4): 97–112.

Shandra, J. M., E. Shor, and B. London. (2008). Debt, structural adjustment, and organic water pollution a cross-national analysis. *Organization & Environment* 21(1): 38–55.

Shaw, D., and T. Newholm. (2002). Voluntary simplicity and the ethics of consumption. *Psychology & Marketing* 19(2): 167–185.

Sheridan, M., C. Moore, and K. Nobbs. (2006). Fast fashion requires fast marketing: The role of category management in fast fashion positioning. *Journal of Fashion Marketing and Management* 10(3): 301–315.

Short, F.-M. C. (2004). Experiment in protecting workers' rights: The garment industry of the US commonwealth of the Northern Mariana Islands. *University of Pennsylvania Journal of Business Law* 7(4): 971–989.

Simpson, L., S. Douglas, and J. Schimmel. (1998). Tween consumers: Catalog clothing purchase behavior. *Adolescence* 33(131): 637–644.

Sit, J., B. Merrilees, and D. Birch. (2003). Entertainment-seeking shopping centre patrons: The missing segments. *International Journal of Retail & Distribution Management* 31(3): 80–94.

Smarzynska, B. K., and S.-J. Wei. (2001). Pollution havens and foreign direct investment: Dirty secret or popular myth? National Bureau of Economic Research Working Paper No. W8465. Cambridge, MA: NBER.

Soron, D. (2010). Sustainability, self-identity and the sociology of consumption. *Sustainable Development* 18(3): 172–181.

Spilková, J., and L. Radová. (2011). The formation of identity in teenage mall microculture: A case study of teenagers in Czech malls. *Sociologický časopis/Czech Sociological Review* 03: 565–586.

Spry, A., R. Pappu, and T. B. Cornwell. (2011). Celebrity endorsement, brand credibility and brand equity. *European Journal of Marketing* 45(6): 882–909.

Staritz, C. (2010). *Making the Cut?: Low-Income Countries and the Global Clothing Value Chain in a Post-Quota and Post-Crisis World*. Washington, DC: World Bank Publications.

Sterling, S., and J. Huckle (eds.). (2014). *Education for Sustainability*. Abingdon, UK: Routledge.

Strizhakova, Y., R. A. Coulter, and L. Price. (2008). Branded products as a passport to global citizenship: Perspectives from developed and developing countries. *Journal of International Marketing* 16(4): 57–85.

Sumner, D. (2003). Implications of the USA Farm Bill of 2002 on Agricultural Trade and Trade Negotiations. *Australian Journal of Agricultural and Resource Economics* 47(1): 99–122.

Sumner, D. (2006). Chapter 10: Reducing cotton subsidies: The DDA cotton initiative. In Anderson, K., and Martin, W. (eds.), *Agricultural Trade Reform and the Doha Development Agenda*. New York: Palgrave Macmillan.

Sumner, D. (2007). U.S. Farm Policy and WTO compliance, AEI agricultural policy series: The 2007 farm bill and beyond. American Enterprise Institute. Available for download at: http://www.aei.org/farmbill.

Taylor, M., and N. Thrift (eds.). (2012). *The Geography of Multinationals: Studies in the Spatial Development and Economic Consequences of Multinational Corporations.* Abingdon, UK: Routledge.

Taylor, S. L., and R. M. Cosenza. (2002). Profiling later aged female teens: Mall shopping behavior and clothing choice. *Journal of Consumer Marketing* 19(5): 393–408.

Techatassanasoontorn, A. A., and R. J. Kauffman. (2005). Is there a global digital divide for digital wireless phone technologies? *Journal of the Association for Information Systems* 6(12): 338–382.

Terasaki, S., and S. Nagasawa. (2012). Celebrities as marketing enhancer case analysis of the alternative food movement and "eco-chic" lifestyle advocacy. In Watada, J., Watanabe, T., Phillips-Wren, G., and Howlett, R. J. (eds.), *Intelligent Decision Technologies* (pp. 263–272). Berlin/Heidelberg: Springer.

Thomas-Hope, E. (ed.). (1998). *Solid Waste Management: Critical Issues for Developing Countries.* Kingston, Jamaica: Canoe Press.

Thompson, N. J., and K. E. Thompson. (2009). Can marketing practice keep up with Europe's ageing population? *European Journal of Marketing* 43(11/12): 1281–1288.

Tokatli, N. (2007). Global sourcing: Insights from the global clothing industry—the case of Zara, a fast fashion retailer. *Journal of Economic Geography* 8(1): 21–38.

Tokatli, N. (2014). 'Made in Italy? Who cares!' Prada's new economic geography. *Geoforum* 54: 1–9.

Tokatli, N., N. Wrigley, and Ö. Kizilgün. (2008). Shifting global supply networks and fast fashion: Made in Turkey for Marks & Spencer. *Global Networks* 8(3): 261–280.

Townsend, J., S. Yeniyurt, and M. B. Talay. (2009). Getting to global: An evolutionary perspective of brand expansion in international markets. *Journal of International Business Studies* 40(4): 539–558.

Toynbee, P. (2003). *Hard Work: Life in Low-Pay Britain.* London: Bloomsbury.

Trindle, J. (2014). GAP gambles on Myanmar. *Foreign Policy Magazine*, July 14.

Tufekci, N., N. Sivri, and I. Toroz. (2007). Pollutants of textile industry wastewater and assessment of its discharge limits by water quality standards. *Turkish Journal of Fisheries and Aquatic Science* 7(2): 97–103.

Tungate, M. (2012). *Fashion Brands: Branding Style from Armani to Zara.* London: Kogan Page Publishers.

Tyler, D., J. Heeley, and T. Bhamra. (2006). Supply chain influences on new product development in fashion clothing. *Journal of Fashion Marketing and Management* 10(3): 316–328.

Um, N.-H. (2008). Exploring the effects of single vs. multiple products and multiple celebrity endorsements. *Journal of Management and Social Sciences* 4(2): 104–114.

Ünay, F. G., and C. Zehir. (2012). Innovation intelligence and entrepreneurship in the fashion industry. *Procedia—Social and Behavioral Sciences* 41: 315–321.

Urdal, H. (2006). A clash of generations? Youth bulges and political violence. *International Studies Quarterly* 50(3): 607–629.

Van Ittersum, K., and N. Wong. (2010). The Lexus or the olive tree? Trading off between global convergence and local divergence. *International Journal of Research in Marketing* 27(2): 107–118.

Varella, M. D. (2014). *Internationalization of Law: Globalization, International Law and Complexity*. Berlin: Springer.

Vergara, S. E., and G. Tchobanoglous. (2012). Municipal solid waste and the environment: A global perspective. *Annual Review of Environment and Resources* 37: 277–309.

Vigneron, F., and L. W. Johnson. (1999). A review and a conceptual framework of prestige-seeking consumer behavior. *Academy of Marketing Science Review* 1(1): 1–15.

Von Moltke, K., and O. Kuik. (1998). Global product chains: Northern consumers, Southern producers, and sustainability. Working Paper #15. Environment and Trade Series Vol. 15. Geneva: United Nations Environment Programme.

Wagner, J., and M. Mokhtari. (2000). The moderating effect of seasonality on household apparel expenditure. *Journal of Consumer Affairs* 34(2): 314–329.

Wagner, U., and C. Timmins. (2009). Agglomeration effects in foreign direct investment and the pollution haven hypothesis. *Environmental Resource Economics* 43(2): 231–256.

Wakelyn, P. J., N. R. Bertoniere, A. D. French, D. P. Thibodeaux, B. A. Triplett, M.-A. Rousselle, W. R. Goynes Jr., et al. (2010). *Cotton Fiber Chemistry and Technology*. Boca Raton: CRC.

Wall, T. M., and R. W. Hanmer. (1987). Biological testing to control toxic water pollutants. *Journal (Water Pollution Control Federation)* 59(1): 7–12.

Walters, D. (2006). Demand chain effectiveness—supply chain efficiencies: A role for enterprise information management. *Journal of Enterprise Information Management* 19(3): 246–261.

Walther, O. (2014). Business, brokers and borders: The structure of West African trade networks. Department of Border Region Studies, Working Paper # 01/14. Sønderborg, Denmark: University of Southern Denmark.

Wang, L. F.-S., and Y.-C. Wang. (2008). Brand proliferation and inter-brand competition: The strategic role of transfer pricing. *Journal of Economic Studies* 35(2): 278–292.

Wang, Y., W. Gilland, and B. Tomlin. (2011). Regulatory trade risk and supply chain strategy. *Production and Operations Management* 20(4): 522–540.

Ward, W. A., M. Bhattarai, and P. Huang. (1999). The new economics of distance: Long term trends in indexes of spatial friction. Working Paper #0299. Clemson, SC: Department of Agricultural and Applied Economics, Clemson University.

Watkins, K. (2002). Cultivating poverty: The impact of US cotton subsidies on Africa. *Oxfam Policy and Practice: Agriculture, Food and Land* 2(1): 82–117.

Wee, T. T. T. (1999). An exploration of a global teenage lifestyle in Asian societies. *Journal of Consumer Marketing* 16(4): 365–375.

Wei, S.-J., and X. Zhang. (2011). Sex ratios, entrepreneurship, and economic growth in the People's Republic of China. Working Paper #16800. Washington, DC: National Bureau of Economic Research.

Weinstein, M. (ed.). (2005). *Globalization: What's New?* New York: Columbia University Press.

Whitelock, J., and F. Fastoso. (2007). Understanding international branding: Defining the domain and reviewing the literature. *International Marketing Review* 24(3): 252–270.

Wigley, S. M., and A.-K. Provelengiou. (2011). Market-facing strategic alliances in the fashion sector. *Journal of Fashion Marketing and Management* 15(2): 141–162.

Wilcox, C., V. Mendez, and C. Buss. (2002). *The Art and Craft of Gianni Versace*. London: Victoria & Albert Publishing.

Willems, K., W. Janssens, G. Swinnen, M. Brengman, S. Streukens, and M. Vancauteren. (2012). From Armani to Zara: Impression formation based on fashion store patronage. *Journal of Business Research* 65(10): 1487–1494.

Williams, K. C., and R. A. Page. (2011). Marketing to the generations. *Journal of Behavioral Studies in Business* 3(1): 37–53.

Winge, T. M. (2008). Green is the new black: Celebrity chic and the "green" commodity fetish. *Fashion Theory: The Journal of Dress, Body & Culture* 12(4): 511–524.

Winters, M. S. (2010). Choosing to target what types of countries get different types of World Bank projects. *World Politics* 62(3): 422–458.

Wolf, M. (2004). *Why Globalization Works*. New Haven, CT: Yale University Press.

Workman, J. E., and S. Cho. (2012). Gender, fashion consumer groups, and shopping orientation. *Family and Consumer Sciences Research Journal* 40(3): 267–283.

Wrigley, N., and M. Lowe. (2007). Introduction: Transnational retail and the global economy. *Journal of Economic Geography* 7(4): 337–340.

Wydick, B., E. Katz, and B. Janet. (2014). Do in-kind transfers damage local markets? The case of TOMS shoe donations in El Salvador. *Journal of Development Effectiveness* 6(3): 1–19.

Yam-Tang, E. P. Y., and R. Y. K. Chan. (1998). Purchasing behaviours and perceptions of environmentally harmful products. *Marketing Intelligence & Planning* 16(6): 356–362.

Yang, D. T., J. Zhang, and S. Zhou. (2011). Why are saving rates so high in China? Working Paper #16771. Washington, DC: National Bureau of Economic Research.

Yang, Y., and M. Mlachila. (2007). The end of textiles quotas: A case study of the impact on Bangladesh. *Journal of Development Studies* 43(4): 675–699.

Yarrow, K., and J. O'Donnell. (2009). *Gen Buy: How Tweens, Teens and Twenty-Somethings are Revolutionizing Retail*. San Francisco, CA: Wiley.

Zakaria, F. (2011). *The Post-American World: Release 2.0 (International Edition)*. New York: WW Norton & Company.

Zhang, B., and J.-H. Kim. (2013). Luxury fashion consumption in China: Factors affecting attitude and purchase intent. *Journal of Retailing and Consumer Services* 20(1): 68–79.

Zhang, Q., M. Zhu, and Y. Yuan. (2014). FDI penetration and manufacturing agglomeration: An analysis based on empirical evidence from 21 industries (2004–2010). *Regional Science Policy & Practice* 6(4): 349–360.

Zhu, N. (2011). Household consumption and personal bankruptcy. *The Journal of Legal Studies* 40(1): 1–37.

Zweifel, T. D. (2006). *International Organizations and Democracy: Accountability, Politics, and Power*. Boulder, CO: Lynne Rienner.

Index

CPSIA information can be obtained
at www.ICGtesting.com
Printed in the USA
BVHW09*1759101018

529426BV00003B/5/P